AQA

Entry Level Certificate in

Science

Gareth Price

Orders: please contact Hachette UK Distribution, Hely Hutchinson Centre, Milton Road, Didcot, Oxfordshire, OX11 7HH. Telephone: +44 (0)1235 827827. Email education@hachette.co.uk Lines are open from 9 a.m. to 5 p.m., Monday to Friday. You can also order through our website: www.hoddereducation.co.uk

First published in 2016 by
Hodder Education,
An Hachette UK Company
Carmelite House
50 Victoria Embankment
London EC4Y 0DZ

Impression number 10 9 8 7 6 5 4

Year 2022

Cover photo © Stocktrek Images, Inc./Alamy Stock Photo
Illustrations by Aptara
Typeset in FS Albert, 13/15 pts, by Aptara Inc.
Printed and bound by CPI Group (UK) Ltd, Croydon, CR0 4YY

A catalogue record for this title is available from the British Library.

ISBN 978 1 4718 7406 2

Contents

Get the most from this book

Welcome to your student book for *AQA Entry Level Certificate in Science*. This book will help you learn more about Biology, Chemistry and Physics and build your practical and communication skills. Some of the key features are shown below.

On your course you will be assessed with a mixture of written tests and practical activities. These activities will be set by your teacher and will make up a portfolio of your work.

You can use this book whether you're studying for Entry Level Single Award or Double Award. This book can also be used to support those studying AQA GCSE Combined Science courses.

✳ Episodes

This book is broken into six parts – 2 for Biology, 2 for Chemistry and 2 for Physics. Each part is split into smaller episodes, so that you can learn the content in manageable chunks.

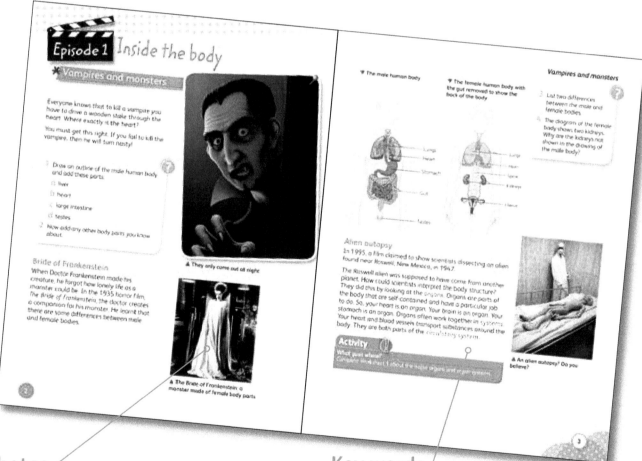

Photos
Look closely at these – they may give you extra information about a topic.

Key words
Important words are shown in blue text. You can find the meaning of these in the Glossary at the back of the book.

Questions

Work through the questions to check you've understood each topic.

You can also use these to practise for assessments.

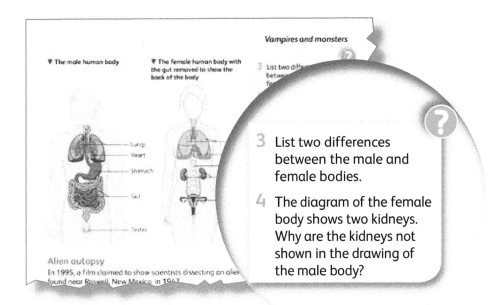

3 List two differences between the male and female bodies.

4 The diagram of the female body shows two kidneys. Why are the kidneys not shown in the drawing of the male body?

Practical

Complete these to develop your practical skills. Your teacher will give you instructions on what to do. You may use this work as part of your portfolio.

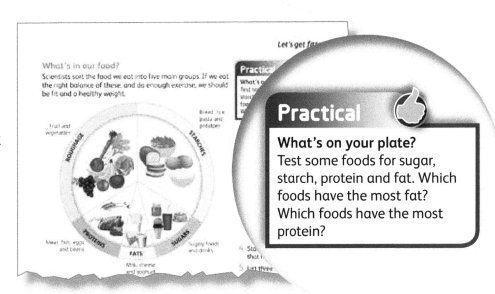

Practical

What's on your plate?
Test some foods for sugar, starch, protein and fat. Which foods have the most fat? Which foods have the most protein?

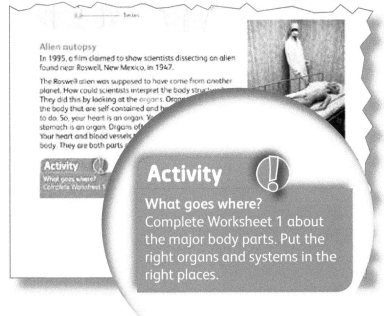

Activity

Try these to see how well you understand a topic and build your investigation skills. Your teacher will give you instructions on what to do.

Activity

What goes where?
Complete Worksheet 1 about the major body parts. Put the right organs and systems in the right places.

✳ Summary

These pages can be found at the end of every section and show a list of things you've learned in class.

How do you feel about each one?

Spend some time thinking about this. It can help you and your teacher to know what you might need extra help with.

 Need more help

 Not sure

 Got it!

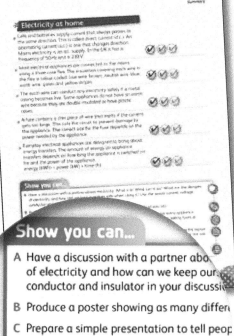

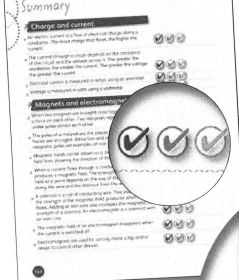

Show you can…

On every Summary page you will find a box of activities that will help you explore the topics you've just learned about. They can be done in class or at home and your teacher will give you ideas of what to do.

You can show what you've learned in lots of different ways.

 Work with a partner or in a group to discuss a topic or share information.

 Show what you know by doing something creative.

 Use your tech skills to share what you've learned.

 Write about what you've learned.

 Get hands-on to build a simple model or device. Be prepared to explain how and why you've built it.

Get the most from this book

Acknowledgements

The publisher would like to thank the following for permission to reproduce copyright material:

AQA material is reproduced by permission of AQA.

Photo credits: p. 2 *t* © Aidar Zeineshev/123RF, *b* ©Moviestore collection Ltd/Alamy Stock Photo; **p. 3** © Luc Novovitch/Alamy Stock Photo; **p. 4** *l* ©Amanaimages/Thinkstock, *r* © Gareth Price; **p. 7** *t* © kasto80/iStock/Thinkstock, *b* © Team It's in the Bag at Grillstock Festival Bristol 2016 - raising awareness of testicular cancer Photo by Chris Cronin; **p. 8** © Gustoimages/Science Photo Library; **p. 10** *t* ©Jani Bryson/iStock/Thinkstock, *b* © Paula Bronstein/Getty Images; **p. 12** © Gastrolab/Science Photo Library; **p. 13** © Arthit Buarapa/123RF; **p. 14** © DigitalVision/Thinkstock; **p. 15** © Nick Savage/ Alamy Stock Photo; **p. 16** © Monkeybusinessimages/iStock/Thinkstock; **p. 20** *tr*© Chris Harvey/123RF, *bl* © sony sivanandan/123RF, *br* © The Science Picture Company/Alamy Stock Photo; **p. 21** © BarnabyChambers/Stock/Thinkstock; **p. 22** © AlexRaths/iStock/Thinkstock; **p. 24** *t* © Fuse/Thinkstock, *b* © Getty Images/iStockphoto/Thinkstock; **p. 26** © Kim Reinick/123RF; **p. 27** © Chlorophylle – Fotolia; **p. 30** *t* © Monkeybusinessimages/iStock/Thinkstock, *b* © Gareth Price; **p. 31** *t* © Liufuyu/iStock/Thinkstock, *bl* © Steven Hobbs/Stocktrek Images/Getty images, *br* © foodfolio/Alamy Stock Photo; **p. 32** *t* © Gareth Price, *b* © Gareth Price; **p. 33** *tl* © E.A. Janes/Age fotostock/Getty Images, *tr* © Gareth Price, *bl* © Gareth Price, *br* © Gareth Price; **p. 34** © Christopher Scott/Alamy Stock Photo; **p. 36** *tr* © Daniel Villeneuve/123RF, *cl* © Wellcome Library, London/Iconographic Collections/wellcomeimages, *bl* © Jim Pruitt/Thinkstock; **p. 38** *t* © Ricardo Beliel/ BrazilPhotos.com/Alamy Stock Photo, *bl* © Gareth Price, *br* © Mshch/iStock/Thinkstock; **p. 39** *tl* © Hxdyl/iStock/Thinkstock, *tr* © Toa55/ iStock/Thinkstock, *cl* © MikaelEriksson/iStock/Thinkstock, *bl* © Gareth Price; **p. 41** © Photofusion/REX/Shutterstock/Rex Features; **p. 43** © Gareth Price; **p. 44** *tl* © Johannes Eisele/AFP/Getty Images, *tr* © Johannes Eisele/AFP/Getty Images, *bl* © ColobusYeti/iStock/Thinkstock, *br* © Tersina Shieh/iStock/Thinkstock; **p. 46** © Ingram Publishing/Thinkstock; **p. 47** © Sebastian Kaulitzki/123RF; **p. 48** © Bennymarty/iStock/Thinkstock; **p. 49** *tl* © Alan64/iStock/Thinkstock, *tr* © Gnagel/iStock/Thinkstock, *bl* © kanoksak Detboon/iStock/Thinkstock, *br* © Imagestate Media (John Foxx)/Amazing Animals Vol 26; **p. 50** *tl* © Ingram Publishing Limited/Ingram Image Library 500-Flowers, *tr* © Imagestate Media (John Foxx)/Seasons V3058, *br* © Ingram Publishing/Thinkstock; **p. 51** © para827/iStock/Thinkstock; **p. 52** © Old Paper Studios/Alamy; **p. 53** © Kjetil Kolbjornsrud/123RF; **p. 56** *tr* © lapas77 – Fotolia, *bl* © Shaiith/iStock/Thinkstock, *br* © Kondratenko/iStock/Thinkstock; **p. 58** *tl* © phil morley/iStock/Thinkstock, *tr* © Rweisswald//iStock/Thinkstock, *bl* © Iaroslav Danylchenko/123RF, *br* © Science Photo Library; **p. 60** *tr* © Lucie Lang/123RF, *bl* © Volodymyr Goinyk/123RF, *bc* © limpido//iStock/Thinkstock, *br* © Anna Ivanova/123RF; **p. 61** © 甘来 /123RF; **p. 62** *tr* © Pictorial Press Ltd/Alamy Stock Photo, *bl* © Agencja Fotograficzna Caro/Alamy Stock Photo; **p. 63** © Francis Apesteguy/ Getty Images; **p. 64** *tr* © Fadil Aziz/Alcibbum Photograph/Corbis, *bl* © Lonely/iStock/Thinkstock; **p. 65** © Olga Khoroshunova/Hemera/Thinkstock; **p. 66** *tl* © Pictorial Press Ltd/Alamy Stock Photo, *tc* © Samir Hussein/WireImage/Getty Images, *tr* © Mim Friday/Alamy Stock Photo, *bl* © Estt/iStock/Thinkstock, *bc* © wisnu haryo yudhanto/123RF, *br* © Blackzheep//iStock/Thinkstock; **p. 70** *tl* © Tyler Boyes/iStockphoto/Thinkstock, *tc* © Bjoern Wylezich/123RF, *tr* © kk – Fotolia, *cl* © AlvaroRT//iStock/Thinkstock, *cc* © Gareth Price, *cr* © Coprid/iStock/Thinkstock, *br* © Gareth Price; **p. 71** *cl* © Peter Macdiarmid/REX/ Shutterstock/Rex Features, *bl* © Pictorial Press Ltd/Alamy Stock Photo, *br* © Gareth Price; **p. 72** *cr* © Gareth Price, *br* © Joe Belanger/Alamy Stock Photo; *tr* © Simon Balson/Alamy Stock Photo, *cl* © Dpa Picture Alliance Archive/Alamy Stock Photo, *cr* © Cai Liang/Alamy Stock Photo; **p. 76** *cl* © Konstantinos Tsakalidis/Alamy Stock Photo, *cr* © Zavalnyuk Sergey123RF, *bl* © Moodboard/Thinkstock; **p. 77** *cl* © Loraks/iStock/Thinkstock, *cr* © Roungchai/iStock/ Thinkstock, *bl* © Dvoevnore/iStock/Thinkstock, *c* © Vladyslav Otsiatsia/iStock/Thinkstock, *b* © Yao Meng Peng/iStock/Thinkstock; **p 79** © Ronald Grant Archive/Topfoto; **p. 80** *t* © szefei/123RF, *bl* © ThamKC/iStock/Thinkstock, *bc* © Dzarek - iStock / Getty Images, *br* © Ratth/iStock/Thinkstock; **p. 81** © Hxdyl/iStock/Thinkstock; **p. 84** © Intellistudies/iStock/Thinkstock; **p. 85** *cl* © Martyn F Chillmaid/ science photo library, *br* © REM118/iStock/Thinkstock; **p. 86** © Andrew Lambert Photography/Science Photo Library; **p. 87** *t* © Exsodus/iStock/Thinkstock, *b* © Javier Trueba/Msf/Science Photo Library; **p. 88** © Chung Sung-Jun/Getty Images; **p. 89** © Copyright 2010 photolibrary.com/ comstock/photolibrary.com; **p. 90** © Antonio_Diaz/iStock/Thinkstock; **p. 91** © Mike Coots; **p. 94** © M. Phillips/WireImage/Gettyimages; **p. 95** *tl* © AlexRaths//iStock/Thinkstock, *tr* © Curraheeshutter/iStock/Thinkstock, *br* © Dejan Kolar/Thinkstock; **p. 97** © X3A Collection/Alamy Stock Photo; **p. 99** © Kevin Frayer/Getty Images; **p. 100** © Mars One/Bryan Versteeg; **p. 102** © overthehill/Fotolia; **p. 103** © Andrew Findlay/Alamy Stock Photo; **p. 104** *tl* © yocamon/iStock/Thinkstock, *tr* © Tibu// iStock/Thinkstock, *bl* © Jevtic/iStock/Thinkstock, *br* © Urban78/iStock/Thinkstock; **p. 105** *t* © Chris2766/iStock/Thinkstock, *b* © Richard Watkins/Alamy Stock Photo; **p. 107** ©Jupiterimages/Creatas/Thinkstock; **p. 110** *tl* © Marvel Studios/Getty Images, *tr* © ScreenProd/Photononstop/Alamy Stock Photo, *bl* © visdia – Fotolia, *br* © John Howard/Photodisc/Thinkstock; **p. 111** *l* © Wavebreakmedia Ltd/Thinkstock, *tr* © yuelan/iStock/Thinkstock, *cr* © Don Farrall/Photodisc/ Gettyimages, *br* © Pearl Bucknall/Alamy Stock Photo; **p. 112** © Yukimasa Hirota/amanaimagesRF/Thinkstock; **p. 113** © Purestock/Thinkstock; **p. 114** © Nagy-Bagoly Arpad/123RF; **p. 115** *t* © Olegkalina/iStock/Thinkstock, *b* © Gareth Price; **p. 116** *bl* © Irina Belousa/123RF, *br* © Jacek Nowak/Alamy Stock Photo; **p. 117** *tl* © Tina Manley/Alamy Stock Photo, *tr* © James Davies/Alamy Stock Photo, *bl* © Inzyx – Fotolia; **p. 119** *t* © Bomboman/iStock/Thinkstock, *b* © NRC; **p. 120** *t* © Tzahiv/iStock/Thinkstock, *bl* © Artjazz/iStock/Thinkstock, *br* © Dennis Hardley/Alamy Stock Photo; **p. 121** *t* © Gareth Price, *b* © inga spence/ Alamy Stock Photo; **p. 124** *l* © Dpa Picture Alliance Archive/Alamy Stock Photo, *r* © 5AM Images/Alamy Stock Photo; **p. 126** *tl* © Jupiterimages/Creatas/ Thinkstock, *tr* © ktsimage/iStock/Thinkstock, *cl* © Killerbayer/iStock/Thinkstock, *cr* © Tokarsky/iStock/Thinkstock, *bl* © Joggiebotma/iStock/Thinkstock, *br* © DigitalVision/Thinkstock; **p. 127** *tr* © IPGGutenbergUKLtd /iStock/Thinkstock, *cl* © Fiorigianluigi /iStock/Thinkstock, *cr* © Minerva Studio /iStock/Thinkstock, *bl* © DigtialStorm/iStock/Thinkstock, *br* © Monchai Tudsamalee/123RF; **p. 128** *tl* ©MartinCParker/iStock/Thinkstock, *tr* © Kneafsey/iStock/Thinkstock; **p. 129** © Visionsport/panoramic/Actionplus/TopFoto; **p. 134** © Everynight Images/Alamy Stock Photo; **p. 135** *tr* © IndiaPicture/Alamy Stock Photo, *cr* © Abraham Adeodatus/123RF; **p. 136** © Thamkc/123RF; **p. 138** © Science & Society Picture Library/Getty Images; **p. 139** *tr* © Olekcii Mach/123RF, *cr* © StockImages/ Alamy Stock Photo; **p. 140** © roberaten/123RF; **p. 141** © Awe Inspiring Images - Fotolia.com; **p. 142** *t* © AF archive/Alamy Stock Photo, *b* © Fouad Saad/123RF; **p. 143** *t* © Esperanza33/iStock/Thinkstock, *b* © Heng Kong Chen/123RF; **p. 146** *tr* © Digital Vision/Thinkstock, *tc* © Mike Dunning/ Dorling Kindersley/Gettyimages, *bl* © Ryan McVay/Thinkstock; **p. 149** *t* © Bloodua/123RF, *b* © Ktsimage/iStock/Thinkstock; **p. 151** *t* © Leonid Serebrennikov/Alamy Stock Photo, *c* © Ingram Publishing/Thinkstock, *b* © Argus - Fotolia.com

t = top. *b* = bottom, *l* = left, *c* = centre, *r* = right

Every effort has been made to trace or contact all copyright holders, but if any have been inadvertently overlooked the Publishers will be pleased to make necessary arrangements at the first opportunity.

Inside the body

✱ Vampires and monsters

Everyone knows that to kill a vampire you have to drive a wooden stake through the heart. Where exactly is the heart?

You must get this right. If you fail to kill the vampire, then he will turn nasty!

1 Draw an outline of the male human body and add these parts:

   a liver

   b heart

   c large intestine

   d testes

2 Now add any other body parts you know about.

▲ They only come out at night

Bride of Frankenstein

When Doctor Frankenstein made his creature, he forgot how lonely life as a monster could be. In the 1935 horror film, *The Bride of Frankenstein*, the doctor creates a companion for his monster. He learnt that there are some differences between male and female bodies.

▲ The Bride of Frankenstein: a monster made of female body parts

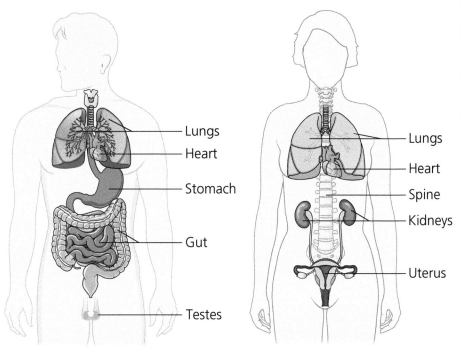

▼ The male human body

▼ The female human body with the gut removed to show the back of the body

- Lungs
- Heart
- Stomach
- Gut
- Testes

- Lungs
- Heart
- Spine
- Kidneys
- Uterus

3 List two differences between the male and female bodies.

4 The diagram of the female body shows two kidneys. Why are the kidneys not shown in the drawing of the male body?

Alien autopsy

In 1995, a film claimed to show scientists dissecting an alien found near Roswell, New Mexico, in 1947.

The Roswell alien was supposed to have come from another planet. How could scientists interpret the body structure? They did this by looking at the organs. Organs are parts of the body that are self-contained and have a particular job to do. So, your heart is an organ. Your brain is an organ. Your stomach is an organ. Organs often work together in systems. Your heart and blood vessels transport substances around the body. They are both parts of the circulatory system.

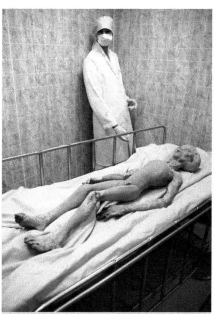

▲ An alien autopsy – Do you believe?

Activity

What goes where?
Complete Worksheet 1 about the major organs and organ systems.

Table 1 Main body systems in humans

System	Which organs does it include?	What job does it do?
Circulatory system	Blood vessels and heart	Transports substances around the body.
Skeletal system	Bones and joints	Holds the body upright. Gives somewhere for the muscles to attach.
Nervous system	Brain and nerves	Controls the body and lets us think.
Digestive system	Mouth and gut	Breaks down food and absorbs it into the body.
Respiratory system	Lungs, nose and throat	Collects oxygen from the air. Passes carbon dioxide out to the air.
Reproductive system	Sex organs (testes and ovaries)	Produces the sex cells (eggs and sperm). Supports growth of fertilised egg.

Transplants

What happens when one of our body organs fails? We cannot build an artificial kidney, lung or liver. The best we can do now is to transplant a healthy organ from a donor.

In 2013, 3000 people in the UK had kidney transplants. Over 6000 were still waiting for a transplant.

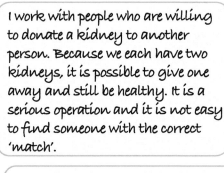

I work with people who are willing to donate a kidney to another person. Because we each have two kidneys, it is possible to give one away and still be healthy. It is a serious operation and it is not easy to find someone with the correct 'match'.

I've had a kidney transplant. It has completely changed my life. I used to have to be connected to a machine to clean my blood three times a week. Now I go to see the doctor once every few months! It's given me the chance to fulfil a lifetime ambition – to visit Tibet! This is me standing on the runway at Lhasa airport!

5 Name the two main parts of the nervous system.

6 Which body system is different in males and females?

7 Write a sentence with the word 'donate' in it.

8 What might make you worry about being an organ donor?

✱ Looking at cells

Organs, like all living things, are made of cells. Cells are so small that scientists need microscopes to see them. A good microscope can magnify the cells almost 2000 times.

The human egg cell is one of the biggest in the body. It is still too small to see without a microscope. If you imagine it was the size of a normal chicken egg, then a human being would be over 1200 m tall. That's bigger than a skyscraper!

Most human cells have the same basic pattern.

▼ A human cell

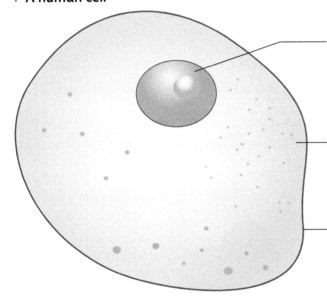

The nucleus: this contains the genes and controls what happens in the cell.

The cytoplasm: most of the cell's chemical reactions take place here.

▲ **If human eggs were the same size as a hen's egg!**

The cell membrane: this controls what gets into and out of the cell.

9 Name the part of the cell that controls it.

10 Explain why all cells need an outer membrane.

Practical

What do cells look like?
Use a microscope to look at cells. Draw what you see. Add labels to your drawing.

Special cells

Although most cells have the same basic pattern, some are specialised to do a particular job.

A group of the same types of cells gathered together is called a tissue. So, muscle tissue is made of muscle cells.

11 Give two differences between a red blood cell and a sperm cell.

12 What job does a nerve cell do?

13 Red blood cells have no nucleus. Explain why this helps them to carry oxygen around the body.

▼ **Specialised cells**

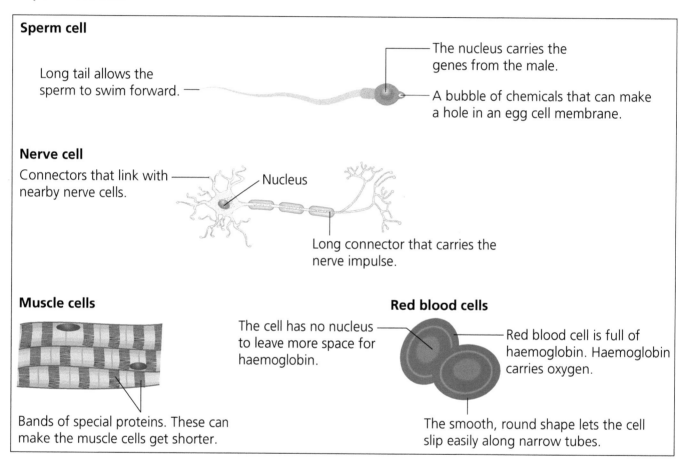

Sperm cell

Long tail allows the sperm to swim forward. —

The nucleus carries the genes from the male.

A bubble of chemicals that can make a hole in an egg cell membrane.

Nerve cell

Connectors that link with nearby nerve cells.

Nucleus

Long connector that carries the nerve impulse.

Muscle cells

Bands of special proteins. These can make the muscle cells get shorter.

Red blood cells

The cell has no nucleus to leave more space for haemoglobin.

Red blood cell is full of haemoglobin. Haemoglobin carries oxygen.

The smooth, round shape lets the cell slip easily along narrow tubes.

Fighting cancer

In a healthy body, the cells work together to do a particular job. One sign of cancer is when the cells divide too quickly and cannot carry out their normal job in a tissue. We can often see this as a lump or tumour.

We often get lumps in our bodies and not all of these are cancers. Tumours can be benign (harmless) or malignant (cancerous). Malignant tumours are different because they can cause other parts of the body to grow out of control as well. So, a cancer may start in the liver but could spread to other parts of the body.

14 Give one way a tissue is different from just a collection of different cells.

Episode 1 Inside the body

Hi! I'm Nathan. I work in a hospital laboratory. I often get cell samples sent to me if a lump has appeared in someone's body. I look at cells taken from the lump to see if they are cancerous. Modern treatments can help many cancer patients beat the disease. People are sometimes still frightened while they wait for me to send through the results though.

15 What can Nathan say to people waiting for their test results to encourage them?

16 How might a family doctor first notice signs of cancer?

It's in the bag!

▲ It's in the Bag is a charity based in Bristol. It campaigns to make young men more aware of testicular cancer.

Every year in the UK, 2200 men get testicular cancer. Sixty of them will die as a result of the disease. Testicular cancer is the commonest cancer in men aged 15 to 45.

Testicular cancer can be cured if it is caught in time. So, how can you tell if you have testicular cancer? You need to check your testes. If they show strange lumps or feel very hard, go to see your doctor.

Almost all lumps in the testes are not cancerous, but you should still go for a check if you find a lump.

17 Why is a strange lump a possible sign of testicular cancer?

18 It is better to get a lump in the testes checked than waiting to see what happens. Why?

Episode 2 Britain's big problem

✳ Let's get fitter

We are getting bigger and bigger in the UK. Too much food and not enough exercise means that we are getting more overweight every year.

People who are very overweight are called obese. Being overweight puts a strain on the heart, the muscles and bones. It can also lead to cancer and diabetes.

People who are underweight are also at risk. They can have less energy and have weaker bones. Their immune system may not work as well, which means they may become sick more easily than other people.

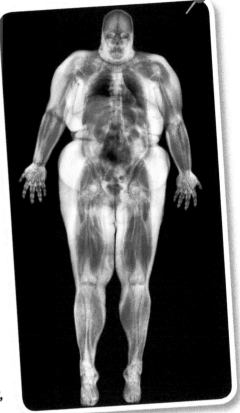

▶ Being overweight puts a strain on the body, as shown in this MRI scan

▼ Which zone do you fit into?

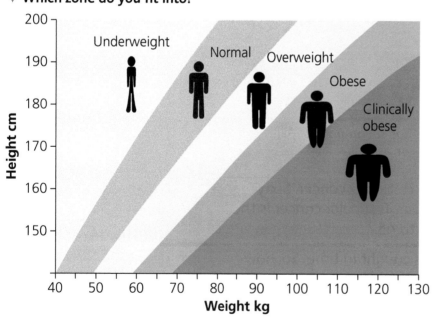

1 Write a sentence with the word 'obese' in it.

2 Chan is 185 cm tall and weighs 62 kg. Look at the graph. What advice would you give him about his weight?

3 Sophie weighs 90 kg and is about 152 cm tall. What are the health risks that she is facing?

What's in our food?

Scientists sort the food we eat into five main groups. If we eat the right balance of these, and do enough exercise, we should be fit and a healthy weight.

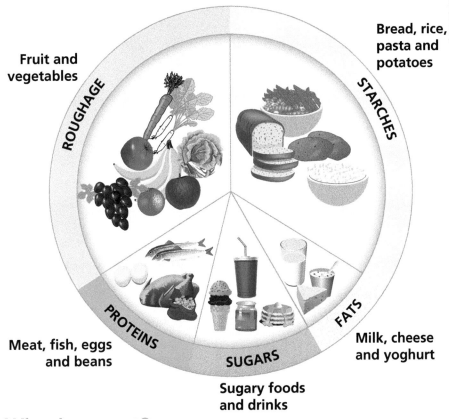

Fruit and vegetables

ROUGHAGE

Bread, rice, pasta and potatoes

STARCHES

Meat, fish, eggs and beans

PROTEINS

SUGARS

Sugary foods and drinks

FATS

Milk, cheese and yoghurt

◀ The main food types and how much of each we should eat

Why do we eat?

The body uses the chemicals in the food to:

✳ give us energy

✳ repair damaged body parts

✳ build new body parts

✳ make a store of energy for a time when food is not available.

4 State two foods you eat that have lots of protein.

5 List three foods that you eat with very little sugar – or no sugar at all!

6 Why should we try not to eat too much fat?

7 Why do weightlifters need to have lots of protein in their diet?

Table 1 Food types, their uses, and what happens if we have too little or too much

Food type	What is it used for in the body?	What happens if we do not get enough?	What happens if we get too much?
Sugars	Energy	We feel tired or weak.	We convert it to fat and put on weight.
Starches	Energy	We feel tired or weak.	We convert it to fat and put on weight.
Fats	Building cells and as a store of energy	We lose weight.	We put on weight. It also seems to damage our blood vessels.
Proteins	Building the body, especially muscles	We can be skinny with weak muscles.	The body converts the spare protein to fat and we put on weight.

The High Energy Biscuit

In 2012, the rains did not come to Mali, Niger, Nigeria and Chad. The crops died and the people had no food. The only chance of survival was to walk many miles to food stations set up by the United Nations.

I sometimes find my work at the food station difficult. I see people who are so close to death. We need to get food into them as soon as possible. We use a HEB – that's a High Energy Biscuit. It is easy to digest, cheap to make and gives them lots of energy. It's amazing how a simple biscuit can save a life!

▲ **One way to deliver your daily biscuit ration – drop them from a helicopter!**

Practical

Biscuit breakdown
The label below shows the ingredients for a High Energy Biscuit. Test some biscuits you like to see how much energy they contain. And do they contain fats, sugars and protein? Could your favourites become rescue biscuits?

Nutrition facts per 100g	
Calories	450 Kcal
Protein	15g
Fat	15g

Ingredients:

Wheat flour, Margarine, Sugar, Soy flour, Sugar syrup, Skimmed Milk powder, Salt, Minerals and Vitamins

Exercise

Playing sports or doing other sorts of exercise can help someone be a healthy weight and avoid illnesses like type 2 diabetes.

People who exercise regularly are usually fitter than people who do little exercise.

Diabetes

Diabetes is a disease that is becoming much more common in the UK. There are two types of diabetes.

Type 1 diabetes is quite rare and is not linked to diet. Type 2 diabetes seems to be linked to diet. Obese people are much more likely to develop type 2 diabetes.

Someone with diabetes does not produce enough insulin. Insulin is a chemical that helps the body to control sugar levels in the blood. Without enough effective insulin blood sugar levels can go out of control.

Table 2 shows what happens when the sugar level goes above or below a safe level.

8 Is exercise fun?

9 Name some sports or exercises that might help someone get fitter.

10 How many times a week do you exercise?

11 a What seems to cause type 2 diabetes?

    b How can you reduce the chances of getting type 2 diabetes?

12 What are the main dangers of type 2 diabetes?

Table 2 What happens when blood sugar level is too low or too high

	Blood sugar level too low	Blood sugar level too high
What are the symptoms?	You feel very weak and tired. In rare cases you can pass into a very deep sleep. This is called a **diabetic coma**.	You do not notice this very much. That is why it is so dangerous!
What is happening inside your body?	Your cells do not have enough sugar to make energy.	Your kidneys, eyes and nerves are damaged. Your blood vessels start to clog with fatty deposits. Many people also develop painful sores on the feet.
How can you deal with this?	A careful diet can help to stop the blood sugar going too low. A spoonful of sugar or jam can help to raise blood sugar quickly and help you come out of a coma.	A careful diet can help to stop the blood sugar going too high. Injections of a chemical called insulin can reduce blood sugar levels.

✱ Break it down

What happens to food once it gets into the body? The gut is a long, muscular tube leading from the mouth to the anus. Muscles in the gut wall squeeze the food along this tube. It can take about 30 hours to get food all the way along the tube.

The digestive system is an important part of the body and contains a number of different organs:

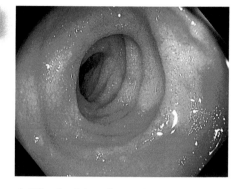

▲ **The inside of a gut seen through an endoscope**

▼ **The human digestive system**

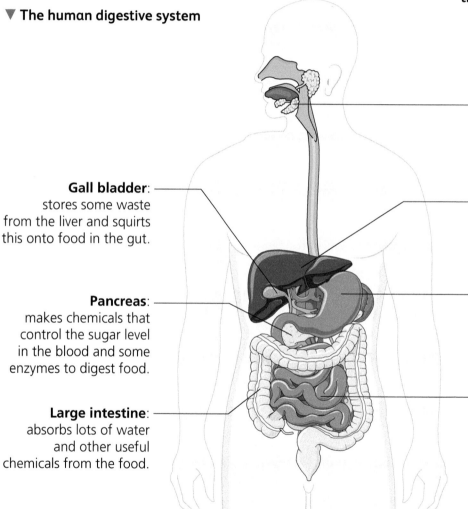

Gall bladder: stores some waste from the liver and squirts this onto food in the gut.

Pancreas: makes chemicals that control the sugar level in the blood and some enzymes to digest food.

Large intestine: absorbs lots of water and other useful chemicals from the food.

Salivary glands: make saliva, which moistens the food and starts to break it down.

Liver: processes food after it has been absorbed into the blood.

Stomach: contains acid and enzymes that break down protein.

Small intestine: produces chemicals that break down food and start to absorb it into the blood.

13 What is the gut wall made from?

14 If you accidentally swallow a coin, how long will it take until it comes out of the anus?

Gastric bands

Doctors can fit a gastric band to help people who are very overweight.

A gastric band squeezes the top of the stomach. When you swallow food it fills the space above the band very quickly. So you do not feel so hungry.

▼ **A gastric band**

Food comes down here.

This band squeezes the stomach here so you feel full more quickly.

Food passes out of the stomach to the small intestine here.

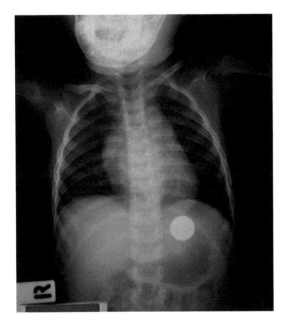

◄ **You were told you not to put coins in your mouth! What happens if you accidently swallow one?**

Digesting dinner

As food passes along the gut, the body produces enzymes and mixes these with the food.

Enzymes can break down the food into substances the body can absorb. These substances pass into the blood and around the body. Each type of food needs its own sort of enzyme.

Practical

Protein digestion
Carry out an investigation to find out how easily an enzyme from the stomach can break down protein.

15 Gastric bands are only offered to people who are very obese. Why not fit them with people who are only slightly overweight?

16 What job do enzymes do in the gut?

13

Episode 3 Going faster

✳ Muscles and marathons

How fast?

Usain Bolt can run 100 metres in under 10 seconds. If he were a car, then he would break the 20 mph speed limit outside many schools in the UK! What makes him such a fast runner?

Muscles

Usain's muscles are just like yours. They are made of protein fibres that can contract. Contract means that they get shorter. His big muscles have lots of protein fibres. If you trained hard, your muscles would develop more fibres.

▲ This female sprinter wants to break the women's 100 m world record. How should she train to do that?

Activity

Who's the strongest?
Are people with longer arms stronger? Or is it the thickness of the muscles that makes the difference? Does it matter if they are male or female? Investigate muscle strength in your class to find out.

1 How could you increase the size of your muscles?

2 Sprinters should eat lots of meat and fish. Why?

Building muscles

3 Draw a graph to show the change in weight of the person in the pictures below.

4 Give two other changes shown in the pictures for Day 0 and Day 360.

Day since training started	Weight kg
0	106
30	97
60	90
90	86
120	82
150	81
180	79
210	78
240	79
270	80
300	83
330	83
360	85

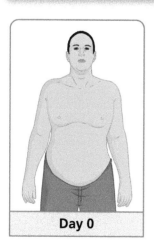

Day 0

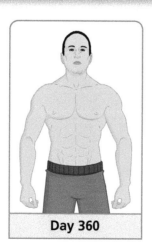

Day 360

Keeping going

Mo Farrah is also a champion. Usain runs 100 m, but Mo runs up to 50 times further! He holds the world record for 5000 m. He finished the race at just under 14 minutes, travelling at an average speed of 13.5 mph.

Mo needs to keep his muscles going for longer than Usain. Mo's training concentrates on building strength in his heart and lungs.

Respiration

The energy to keep athletes running comes from their food. The body converts food into energy by respiration.

Respiration is a chemical process that happens in every cell in the body. The word equation below shows what happens.

glucose + oxygen → carbon dioxide + water + energy

Glucose is a type of sugar and comes from food.

* Our lungs collect the gas oxygen from the air and load it into the blood.

* The blood carries it to the muscles.

* The muscles use the oxygen and sugar to make energy.

The muscles also produce carbon dioxide and water. The carbon dioxide passes back into the blood and goes back to the lungs where it is breathed out. The heart keeps the blood moving around the body.

▲ If an athlete wants to run like Mo what training should they do?

▼ Oxygen gets to muscles and carbon dioxide gets back to the lungs

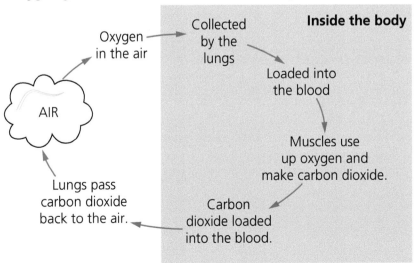

5 Give two differences that you can see between the body shapes of sprinters and marathon runners.

6 Why does the blood have to keep moving around the body?

7 Cigarette smoke does not affect muscles very much. However, very few top athletes smoke. Why?

✱ Pump it up!

Blood donors

▲ **Giving blood could save a life.**

You have about 5 litres of blood in your body. Blood donors give about 470 ml at each donation session.

This blood is given to people who need it because they have been hurt or are ill. Some chemicals extracted from blood make powerful drugs.

What goes where?

Blood is a way to transport things around the body. Table 1 shows the most important substances that blood takes around the body. It also carries many other substances.

If you do not have enough blood, then this transport slows down. You can feel tired and ill.

Table 1 Substances carried by the blood

Substance	Carried from	Carried to
Oxygen	The lungs	All cells to use in respiration.
Carbon dioxide	All cells	The lungs to be breathed out.
Food substances	The gut	All cells to be used for energy and to build new cells.
Waste substances	All cells	The kidneys to pass out in urine.

8 When you turn 18, you may be able to give blood. Would you? Why? Why not?

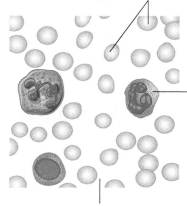

Red blood cells: these carry oxygen around the body.

White blood cell: there are many types of these. They all help to protect the body against disease.

Plasma: this is the liquid the blood cells are carried in.

▲ **What is in a sample of blood?**

9 What do red blood cells do?

10 What is carried from the lungs to every cell in the body?

The circulatory system

The heart is a pump that is made of strong muscle. It is part of the circulatory system. It pumps blood around the body every second of every day.

This blood carries food and oxygen to the cells of the body and carries carbon dioxide and other waste substances away.

One pump or two?

The heart is actually two pumps stuck together to form a double circulatory system.

The right side pumps blood from the body to the lungs. The blood then goes back to the heart and the left-hand pump pushes it all the way around the body.

▼ **The two sides of the heart**

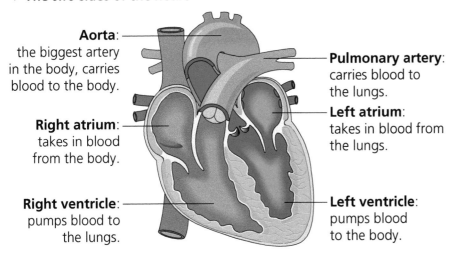

Aorta: the biggest artery in the body, carries blood to the body.

Pulmonary artery: carries blood to the lungs.

Right atrium: takes in blood from the body.

Left atrium: takes in blood from the lungs.

Right ventricle: pumps blood to the lungs.

Left ventricle: pumps blood to the body.

▼ **The circulatory system**

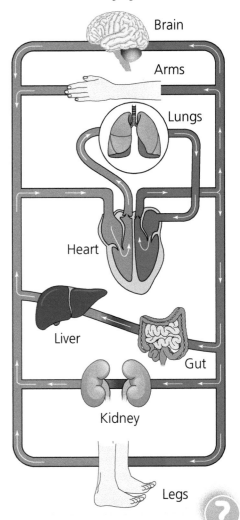

Brain

Arms

Lungs

Heart

Liver

Gut

Kidney

Legs

Practical

Working it
Your muscles need more oxygen and glucose when they work harder. How does this affect your breathing and heart rate? Carry out an investigation to find out.

11 What is the heart wall made of?

12 How can you tell if the heart is still beating?

13 Give the name of the large artery that takes blood from the heart to the rest of the body.

14 Which side of the heart is bigger? Why do you think this is?

15 Give two differences between blood going towards the lungs and blood coming away from the lungs.

Summary

✳ Cells and tissues

» All living things are made of cells. These are so small they can only be seen through a microscope.

» Cells contain a nucleus (controls the cell and contains the genes), cytoplasm (chemical reactions take place here) and a cell membrane (controls what passes in and out).

» Many cells are specialised to do a particular job.

» A tissue is a group of the same type of cells.

✳ Organs and systems

» Organs usually do one job in the body. The heart and brain are organs.

» Systems are groups of organs that work together to do a large job in the body. The circulatory system consists of the heart and blood vessels.

» The heart is made of muscle and has two pumps. One pumps blood to the lungs and the other pumps blood to the body.

» Enzymes speed up reactions in living things. The gut uses enzymes to break down large, complex food molecules into smaller, simpler ones.

✳ Food and feeding

» Most foods are mixtures of proteins, carbohydrates and fats.

» A balanced diet includes all the major food groups, vitamins and minerals, roughage and water.

» The gut is a long muscular tube that runs from the mouth to the anus. It produces enzymes that break down large complex food molecules into smaller simpler ones. These small molecules can be absorbed by the body.

✱ Fitness and health

» Respiration releases the energy needed for living processes. It is represented by the equation:

glucose + oxygen → carbon dioxide + water (+ energy)

» Lifestyle has an effect on people's health. People who exercise regularly are usually fitter than people who take little exercise. People who eat a balanced diet and take plenty of exercise tend to be fitter.

» A healthy diet contains the right balance of the different foods you need and the right amount of energy.

» When the body has to work harder, the heart beats faster so the pulse rate rises. You also breathe more deeply to take in more oxygen and get rid of more carbon dioxide.

» Food is used to produce energy for activity. Any spare energy is used to produce fat as a store to energy.

» People who carry too much body fat are called obese. Obesity can lead to an increased risk of diabetes, some types of cancer, and puts a strain on the heart and skeleton.

Show you can...

A Use the clue cards to explain to a partner the ways that red blood cells, nerve cells and sperm cells are adapted to their function.

B Get a partner to draw round your body and make a giant poster showing the different parts inside your body.

C Create a presentation showing how to carry out tests for starch, sugar, fat and protein. Include a voiceover if you can.

D Prepare a simple leaflet for a gym to explain how exercise can benefit people who join the gym. Describe the exercises you should do to: gain muscle, improve your general health or lose weight.

E Create a model cell for a museum exhibit. The exhibit should be suitable for 11-year-old visitors. Add labels and prepare a short quiz for visitors to complete when they have seen the exhibit.

Episode 4 Diseases

✳ Attack!

Dateline Oct 10, 2015. Zombie invasion!

In October 2015, London was invaded! Thousands of people dressed as zombies marched through London raising money for charity.

Zombies are the undead and almost impossible to kill! One way to become a zombie is to be infected with a mysterious virus.

Infections

Many illnesses (not zombies!) are caused by tiny living organisms that get into the body.

Organisms that cause disease are called pathogens. There are many types of pathogen. Two of the most important groups are bacteria and viruses.

▲ A zombie from London

Bacteria	Viruses
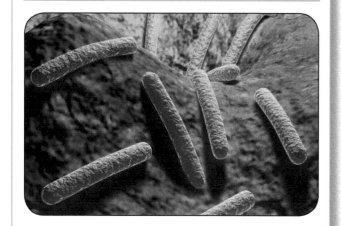	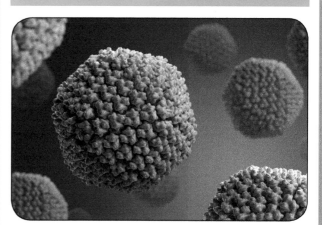
Many bacteria are helpful. Some turn milk into yoghurt or break down wastes in the soil.	All are harmful but only some cause disease in humans.
100 times smaller than a human cell.	10 000 times smaller than a human cell.
Antibiotics work against them.	Antibiotics do not work against them.
Can cause: Sore throat and earache, food poisoning and meningitis.	Can cause: Flu, AIDS, mumps, measles, rubella.

Practical

Killing microorganisms
Bleach kills germs. How strong does it have to be to do this? Investigate how well bleach can kill microorganisms when you make it very dilute.

1 Give one difference between viruses and bacteria.

2 What do you think would happen if we were able to kill all the bacteria in the world?

SARS

SARS is a disease caused by a virus. It makes you feel weak, you ache all over and get a high temperature. For every ten people who catch SARS, one will die!

In 2003 an outbreak started in southern China. To try to prevent the disease spreading, people began to wear face masks. People entering Hong Kong from mainland China were tested for high temperatures, a sign of the disease. If your temperature was too high, then you might not have been allowed into the country.

The plans worked and, by March 2004, the SARS epidemic had passed.

The diagram below shows just how a virus works inside the body to make us ill.

▲ Some people wore face masks to protect against SARS in the 2003 outbreak

▼ **How viruses make us ill**

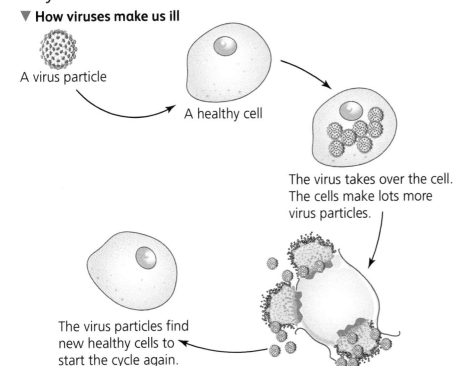

A virus particle

A healthy cell

The virus takes over the cell. The cells make lots more virus particles.

The virus particles find new healthy cells to start the cycle again.

The cell bursts. The virus particles escape. Some parts of the virus are poisonous.

3 How do you think the SARS virus spreads? (The masks give you a clue.)

4 Doctors did not give antibiotics to people with SARS. Why?

Flu jab?

Most years, we have an outbreak of influenza in the UK. We call it flu. Flu is caused by a virus. People with flu get aches and pains, feel weak and will have a high temperature.

You can get a flu jab. This helps to protect against the disease. A flu jab is a vaccination. It contains chemicals from the virus that trigger your body's defence reaction.

White blood cells in the body produce chemicals called antibodies. Antibodies destroy the virus before it can do any damage. The body's self-defence system is called the immune system.

Vaccinations increase our natural immunity. The first time the body produces antibodies, it is quite slow. The next time the virus gets into the body, the immune system produces lots of antibodies very quickly.

▲ A flu jab can help protect against the disease

▼ **How the body responds to a vaccination**

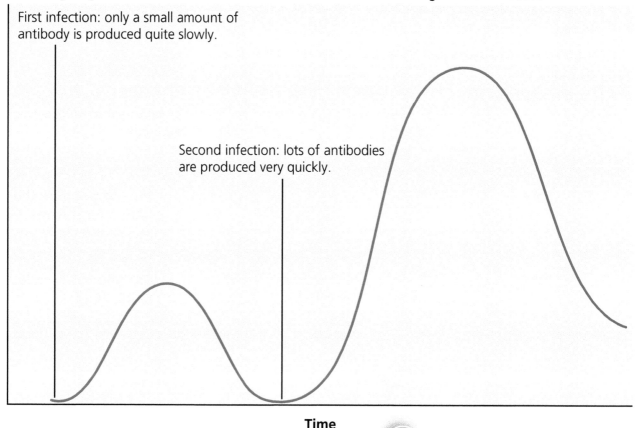

First infection: only a small amount of antibody is produced quite slowly.

Second infection: lots of antibodies are produced very quickly.

Amount of antibody in blood. This protects the body.

Time

5 What vaccinations have you had?

6 What diseases do these protect against?

Treating illnesses

Sometimes, when we are ill we just need to rest and wait. Sometimes we take drugs to destroy the pathogens (the microorganisms that are making us ill).

Antibiotics are drugs that destroy bacteria. There are many different types of antibiotics. The most famous is probably penicillin.

All drugs need to be tested before we use them. Some drugs can kill pathogens but also harm the patient.

Side effects are things the drugs do that we do not want. Side effects include feeling sick, having a high temperature or even an increased risk of serious illness. People may become addicted to the drugs they are taking.

Drug trials are often double blind. The researcher provides two sets of tablets. One set contains the drug to test. The other set does not. Even the researcher who gives the patients the drugs does not know which is which, to stop them influencing the patients. The tablets without the real drug are the control or placebo.

▼ **How drugs are tested**

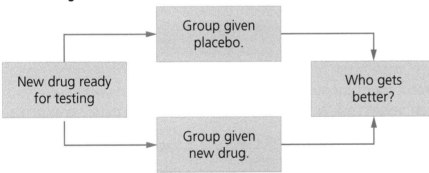

A group of patients is divided into two. One set are given the real drug. The other set is given the placebo. The patients do not know whether they are getting the real drug or the placebo. The doctors giving them the drug do not know. Only the researcher knows.

If the people who get the real drug get better and those given the placebo drugs do not, then the researcher knows the drug works. If both sets of patients get better, then the researcher knows the drug has no effect.

7 Why are drugs trials 'double blind' and not just 'blind'?

8 Imagine you had a very serious illness. A researcher claims they have a drug that will help you, but it has not been tested for safety. We do not even know that it will work. What would you do? Why?

✳ Danger!

▲ Watching scary films produces adrenalin in the body

Scary stuff!

What happens when you get scared? Your hair stands on end. You begin to sweat. You might even begin to feel sick.

All these changes are controlled by adrenalin. Adrenalin is a hormone (a chemical made in your body). It gets your body ready for action. You may begin to feel 'on edge' or 'twitchy'. Some sports people think they need to be a bit scared before important competitions, just to help them do their best.

Endocrine glands make hormones and pass them into the bloodstream. Adrenalin is made by the adrenal glands.

John carries adrenalin around with him. He is allergic to peanuts. If he eats them, or even just smells them, he can go into shock. His heart could stop. A shot of adrenalin from a specially designed injector could keep his heart going and save his life.

?

1 What types of things make you nervous?

2 Do you think being nervous before a match or a test helps you do better? Why? How could you find out?

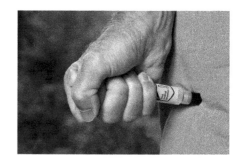

▲ John is giving himself a shot of adrenalin with an epi-pen, which he carries with him

The blood carries the hormones to where they are needed. These are called the hormone's target organ. Adrenalin affects the heart, brain, skin, eyes, and even your gut!

▼ **Where hormones are produced and used**

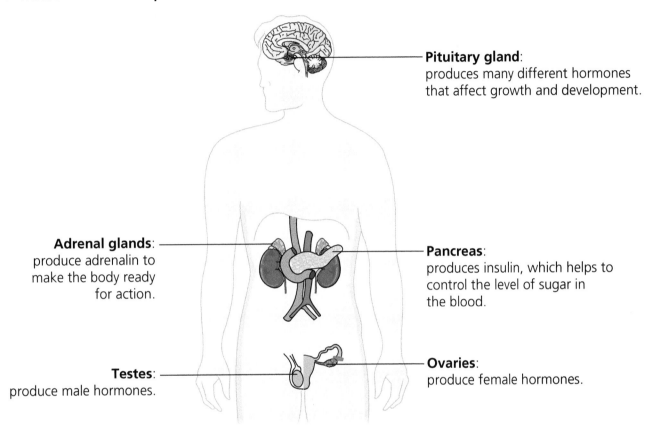

Pituitary gland: produces many different hormones that affect growth and development.

Adrenal glands: produce adrenalin to make the body ready for action.

Pancreas: produces insulin, which helps to control the level of sugar in the blood.

Testes: produce male hormones.

Ovaries: produce female hormones.

The sex hormones

Puberty is the time when our sex organs begin to start developing. In boys the testicles start to produce sperm. In girls, the ovaries start to release eggs. These changes are controlled by hormones released from the brain.

Hormones also make other changes happen in our bodies. These mean men tend to be taller, have more hair on their body and face and develop deeper voices than women. Women develop breasts during puberty.

3 What hormone does the adrenal glands produce?

4 What part of the body makes the hormone insulin?

5 What happens if you do not have enough growth hormone?

The sex cycle

Males produce sperm continuously throughout their adult life. Females tend to produce one egg a month.

If the egg is fertilised, then it will develop into a baby inside the body. Hormones control the release of an egg and they make the lining of the uterus develop. The uterus is the space where the baby will develop. If the egg is fertilised by a sperm, it will sink into the wall of the uterus and grow. This baby will be born about 9 months later.

If the egg is not fertilised, it is passed out of the body with the lining of the uterus and some blood. This is known as a period or menstruation.

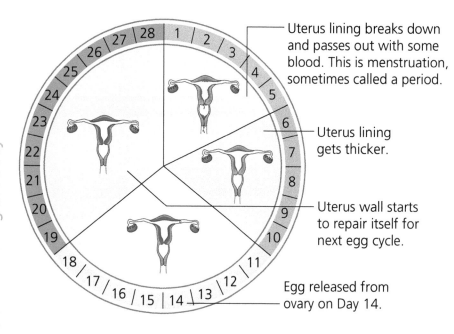

Uterus lining breaks down and passes out with some blood. This is menstruation, sometimes called a period.

Uterus lining gets thicker.

Uterus wall starts to repair itself for next egg cycle.

Egg released from ovary on Day 14.

6 What happens if the egg is fertilised?

7 Why does the uterus need a good blood supply?

◀ **The menstrual cycle**

Controlling fertility

The contraceptive pill is the commonest medicine in the world. Millions of doses are produced every year. It contains female hormones that fool the body into thinking it is pregnant. This hormone stops the ovaries from making and releasing an egg. It gives women more control over when they have children. If they stop taking the pill, they can usually become pregnant normally afterwards.

Some hormones encourage the body to produce eggs. Doctors prescribe these drugs in carefully measured doses. Some of the early fertility treatments led to women producing too many eggs. Twins and triplets were common! Sometimes men are given male hormone treatments to encourage sperm production.

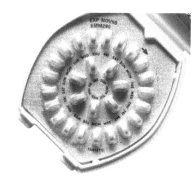

▲ **Contraceptives give people choices about whether to have children**

8 What are the advantages of being able to control fertility

9 Give one possible problem if people control their fertility with drugs.

Reflexes

Hormones like adrenalin take a few seconds to work. When the body is in immediate danger, it needs a faster response.

When you pick up a hot object you drop it very quickly. Before you realise you have been burnt! This is an example of a reflex.

Reflexes are very, very fast reactions that protect the body. We do not need to learn reflexes. We are born with them.

A video game champion will react much more quickly than a beginner. These reactions are learnt. Sometimes people call these very fast reactions reflexes.

10 What is the advantage of a reflex like blinking when dust gets into your eyes?

Activity

Lightning speed
Carry out an investigation to see how well two people can do something very quickly. Perhaps they could catch a ball or get through a video game level. Do these sorts of very fast reaction improve with practice?

▲ You need to practise if you want to get past the big boss on Level 7!

Summary

✱ Illness and disease

» Infectious diseases are caused by microorganisms that get into the body. These are called pathogens. Pathogens cause disease by releasing poisons or destroying cells.

» You can catch infectious diseases from other people. You become infected.

» Viruses are very small pathogens that cause diseases like AIDS, SARS, flu and measles.

» Bacteria are pathogens that cause diseases like sore throats, earache and meningitis.

✱ Defending the body

» The body's defence system is called the immune system. It has two parts: white blood cells and antibodies.

» White blood cells swallow pathogens and break them down. White blood cells also produce chemicals called antibodies, which destroy pathogens. Each pathogen has its own particular antibody.

» A rise in temperature, called a fever, is often a sign of infection.

» A vaccine contains small quantities of dead, specially treated viruses or parts of a pathogen. When it is injected into the body, it stimulates the white blood cells to produce antibodies. The next time the same pathogen enters the body, antibodies are produced rapidly.

✱ Hormones

» Hormones are chemicals that are produced by endocrine glands. Hormones do not affect the gland where they are made. They pass straight into the bloodstream.

» Hormones travel around the body in the bloodstream to reach their target organs. When they reach their target organs they change what the organs are doing.

» Hormones can have effects across the body. Changes in male and female bodies at puberty are caused by hormones.

» Some hormones in females work together to produce the menstrual cycle. This means women produce an egg and have a period roughly once a month between the ages of about 11 and 55.

» A package of female hormones is used in the contraceptive pill to stop women producing eggs.

» A different set of female hormones can encourage women to produce more eggs. These hormones are used in fertility pills. Some male hormones can help men to produce more sperm.

✱ Reflexes

» Reflexes are automatic responses that happen very quickly. Reflexes usually protect the body from danger.

» The blinking reflex protects our eyes from dust and helps to wash them clean. We pull our hands back from hot objects very quickly to protect our skin against burning. This is also a reflex.

» Some people describe any very fast movement as a reflex, e.g. catching a ball or playing a very fast video game.

Show you can...

A Imagine you are a doctor. Your partner should act as a patient with bad flu. Explain why you are not going to give them antibiotics.

B Produce a simple poster to show the main symptoms of flu. Include advice about what to do if you get flu.

C Produce a presentation showing how to test a bleach to see how well it works against microorganisms.

D Produce a leaflet to advise people who are planning a round-the-world trip. What vaccinations do they need? How can they protect their health when they are travelling?

E Researchers often have to handle dangerous microorganisms. Build a simple face mask that still lets you breathe but keeps out dust and germs. Explain how you could test it to see if it kept out microorganisms.

Episode 1

Biology 2

Up the garden path

✶ Welcome to Rooftops!

> Five years ago some people wanted to knock this building down and build a new car park. We decided homes for people were more important than cars. We converted the flat roof into a garden. We grow tomatoes, courgettes, lettuce, sweetcorn, carrots and even flowers for the residents.

Sunlight

Green plants need sunlight to make sugar (glucose) from water and carbon dioxide. The plant also gives out oxygen as a waste product. This is called photosynthesis.

▼ **Photosynthesis in a leaf**

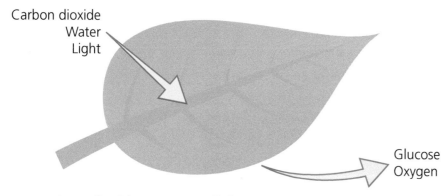

Carbon dioxide
Water
Light

Glucose
Oxygen

carbon dioxide + water + light ⟶ glucose + oxygen

1 What are the advantages of a roof garden?

2 What problems are there with a roof garden?

3 Why is oxygen useful to animals?

4 The drink made from this sugar cane tastes sweet. Why?

▲ This machine crushes the sugar cane to release the sugary sap. This makes a great drink on a hot day

Activity

Growing in the city
Many cities have coloured lights blazing all night. Does this mean that the plants grow all the time?

Plan an investigation to find out what sort of light is best for plants.

Water

If we are ever going to live on Mars, we will need to grow our own food there. There is no soil on Mars. Perhaps we can grow plants in water or water and Mars-dust mixtures?

 5 Why do plants need water?

▲ Could this be used to grow plants on Mars?

Practical

Algae for lunch
Algae are water plants that do not need soil to grow. We can even eat them!

How can we grow the most algae to eat? Find out how the light level affects the growth of algae.

▲ Laverbread: it's black and tastes like salty water and cabbage. You either love it or hate it!

Space to grow

What do you need to grow a plant? Just some sunlight, water and space to grow. Sometimes you don't even need soil!

▲ This field of lettuce plants in China is completely weed-free!

▲ Unfortunately this garden in Leicester is not!

Why do farmers work so hard to keep their fields free of weeds? When plants grow close together they compete for sunlight, water and space. If water is in short supply, then some plants will wilt.

6 What are the plants in the weedy garden competing for?

Activity

Growing more in the same space
Investigate the effect of planting lots of lettuce plants close together. Do you get more food, or just lots of small plants?

The rat race

Animals also compete with each other. There are thousands of rats in London. They live in the London Underground, bins, alleyways and derelict buildings. They depend on the food dropped by people.

If there is lots of food, competition is low and the rats multiply quickly. If the amount of food available goes down, competition increases and some rats may go hungry.

7 What do the rats compete for in London?

8 What happens to the number of rats in a city if rubbish is not collected? Why?

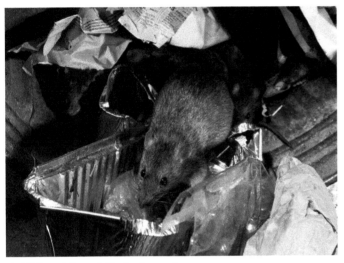

▲ This rat is not likely to go hungry

▲ There is plenty of water for the plants in this Cambodian floating village, What might the plants in the pots be competing for?

▲ What is in short supply for these Australian plants?

▲ What are the plants in these fields in China competing for?

✳ Food chains

Cecil the lion was killed by a dentist from Minnesota, USA in 2013. The dentist used a bow and arrow and took away the skin and head as trophies.

Every year about 50 lions are killed for trophies in Zimbabwe. Most of them are killed by foreign hunters.

Hunters have been going to Africa for many years to kill animals and then take away parts for souvenirs.

▲ When Cecil was killed, it caused a lot of comment on social media

9 Why do hunters want to kill lions?

10 What are the advantages to local people in Zimbabwe of foreign hunters?

11 Do you think it's right to kill animals even when we don't need them for food?

Food from the Sun

All of our food comes from the Sun. Sunlight makes plants grow. Some animals eat these plants to get this energy. They are called herbivores.

Some animals eat other animals. They are called carnivores.

Scientists use a diagram called a food chain to show what an animal eats. To show that a slug eats a lettuce, we draw an arrow from the lettuce to the slug. The lettuce is called a producer and the slug is a consumer. All food chains start with a producer.

▼ A simple food chain

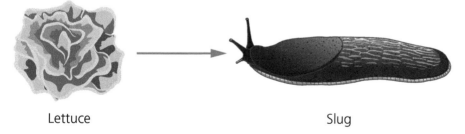

Lettuce   Slug

We can add more links to this chain. A bird like a thrush could eat the slug. A bird of prey like a hawk could eat the thrush. Food chains do not go on forever. Food chains never have more than five links. They all end in something that is not eaten by anything else. This is called the top carnivore.

▼ **A longer food chain**

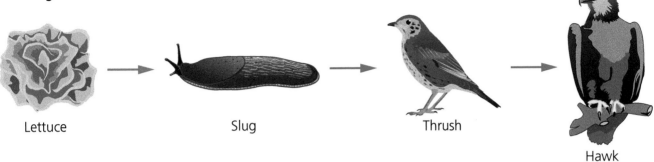

Lettuce   Slug   Thrush   Hawk

The food chains in an area connect together to make a food web. This is because many living things are part of more than one food chain.

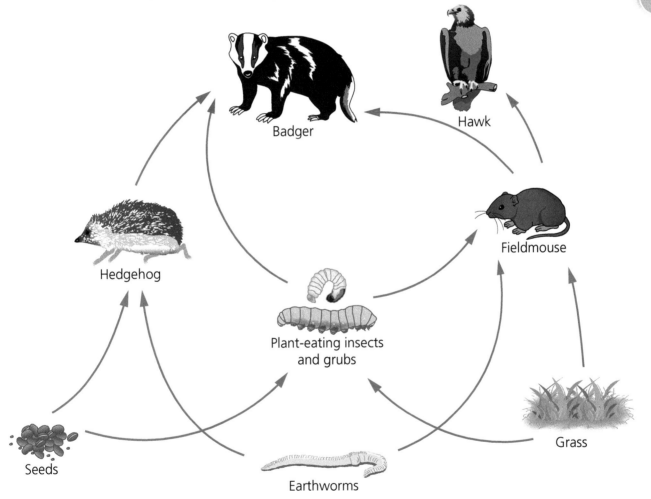

Badger   Hawk

Hedgehog

Fieldmouse

Plant-eating insects
and grubs

Grass

Seeds

Earthworms

12 Which things in this food chain would you call producers?

13 Which animals are carnivores?

14 Which animals are the top carnivores?

15 What would happen to the hawks if all the fieldmice were killed by a disease?

Episode 2 Death and decay

✳ Cheating death?

Maori warriors, Chairman Mao, Tutankhamen and Doris Willis of Wyoming all have something in common. Their bodies were preserved after they died. For the Maori warriors it is only the head. For Doris it was the whole body.

▼ Horatio Robley with the severed heads of Maori warriors

▲ Doris's body is stored in liquid nitrogen at –200 °C

Decay and decomposition

Living things decay when they die. Scientists say they decompose. The chemicals in their body break down to make simpler chemicals. These chemicals are used by living things to grow.

1 Doris will be defrosted in hundreds of years' time. She hopes that by then science will be clever enough to bring her back from the dead. Would you like this to happen to you? Why? Why not?

Practical 👍

Decomposing grass
How hot will a pile of grass clippings get as they decompose? Design an investigation to find out.

▲ Grass cuttings do not last forever. Microorganisms decompose the grass cuttings to make compost. The microorganisms produce heat as they digest the grass.

Recycling living things

Microorganisms help to break down dead bodies. They are also able to break down other wastes produced by living things. We're talking faeces here!

▼ **This diagram shows how living things are involved in recycling chemicals**

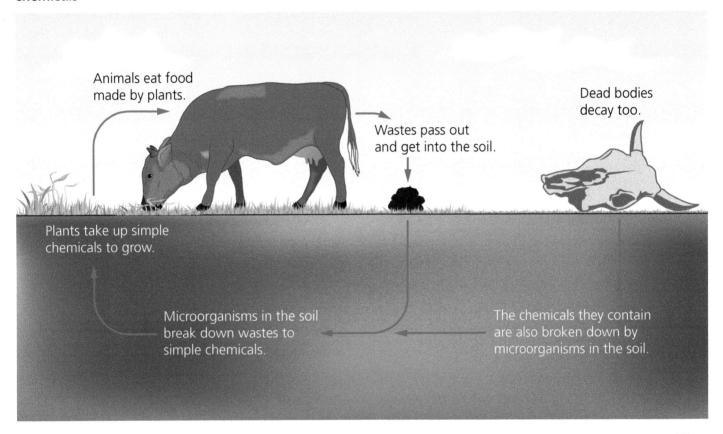

Animals eat food made by plants.

Wastes pass out and get into the soil.

Dead bodies decay too.

Plants take up simple chemicals to grow.

Microorganisms in the soil break down wastes to simple chemicals.

The chemicals they contain are also broken down by microorganisms in the soil.

Cows eat grass and make milk for their owners.

Cows also make faeces, which rots down in the soil and releases good fertilisers for the growing crops!

2 Why do farmers put manure on their fields?

3 The best manure is not fresh and steaming! It is good to store it for a year and then add it to the soil. Why?

Practical

Well preserved
What affects how quickly something decomposes? Plan an investigation into how quickly slices of apple go mouldy.

Episode 3 Safe Earth

✳ Let's save the world!

▲ The people on this boat are working together to try to stop activities that they think damage the planet

We are all told to save the planet. But what from? What is damaging the planet? Is it us?

Human beings produce lots of waste just because they are alive. What you flush down the toilet has to go somewhere.

But we also make waste when we make cars or build smartphones or travel to China or even grow wheat for our breakfast cereal. Many of these wastes are dangerous unless they are treated properly.

A pollutant is a substance created by human activity that gets into the environment and can do some damage.

▲ Traffic jams – Bangkok style!

1 List any things that you do to try to 'save the planet'. This could include recycling or turning off the lights when you do not need them.

2 Explain how each of the things you listed in question 1 helps the planet.

▲ Pollutants are being released from these power station chimneys

▲ Where did all this waste come from?

▲ Everything we flush away must end up somewhere

▲ Leaking oil drums in Tibet

▲ Even shopping can create waste

Practical

Poisoning the plants
Abandoned mines or heaps of waste from lead mining often still contain small amounts of the metal. Investigate how lead dissolved in water can affect the growth of plants.

3 What pollutants can you see in the photographs on these pages?

Acid rain

▼ How acid rain is formed

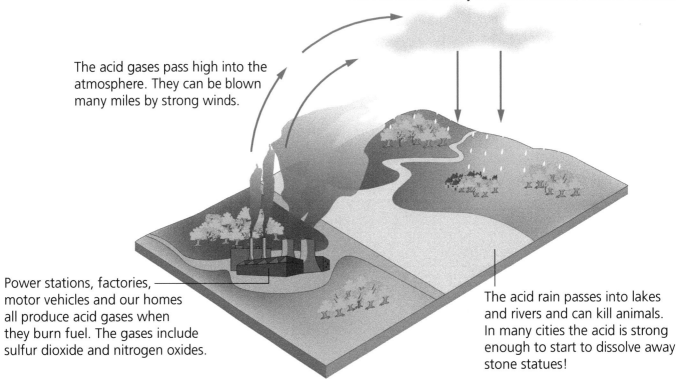

The acid gases can dissolve in rain or even snow. This falls from the sky and into streams, rivers and lakes.

The acid gases pass high into the atmosphere. They can be blown many miles by strong winds.

Power stations, factories, motor vehicles and our homes all produce acid gases when they burn fuel. The gases include sulfur dioxide and nitrogen oxides.

The acid rain passes into lakes and rivers and can kill animals. In many cities the acid is strong enough to start to dissolve away stone statues!

Sewage

There are many sources of sewage that can pollute different areas. Sewage is everything you flush down the toilet and everything that goes down the plughole. It is taken to sewage treatment works and is converted to water, carbon dioxide and dissolved minerals. Sometimes the sewage works are overloaded and sewage is released straight into the rivers and seas.

In rivers, the sewage makes certain microorganisms grow too much. These use up all the oxygen in the water and kill the fish in the river. Fertilisers and animal waste from farms can have the same effect.

4 a What two things produce most of the sulfur dioxide in the air?

 b How could you reduce the amount of sulfur dioxide that is produced?

 c How does the sulfur dioxide come back to the ground?

5 What damage does acid rain do to:

 a buildings

 b living things?

▼ Where does sewage come from and where does it go?

From our homes and workplaces

Washing and bathing water | Water and faeces from the toilet | Waste water from the washing machine or kitchen sink

Enters the drains and sewers

↓

Treated at sewage works and released to rivers, and seas. If too much sewage is produced it is released before fully treated.

From farms

Manure spread onto fields.

Rain washes chemicals from the manure into streams, rivers and lakes.

▲ In the summer many more people go to Cornwall for holidays. Sometimes they overload the sewage works. The sewage is then pumped out to sea.
Surfing anyone?

Damage? What damage?

Table 1 Pollutants and damage that they can do

Pollutant	Where does it come from?	Where does it go?	What damage can it do?
Raw sewage	Human beings	Into streams, rivers, seas and oceans	It helps water plants and microorganisms to grow. These can poison the river and use up all the oxygen. Everything in the rivers dies.
Fertilisers from farms	Farm fields		
Toxic chemicals	Factories and mines	Into soil, streams, rivers, seas and oceans	Many are poisonous, for example, lead can damage our brains.
Acid gases like sulfur dioxide, nitrogen oxides	Burning petrol, diesel, coal and oil	Into the atmosphere	Causes acid rain, which can destroy buildings and poison lakes and rivers.
Chemicals used to kill pests or weeds on farms	Farms and gardens	Into the soil, streams, rivers, lakes and oceans	These chemicals are poisons that can cause damage long after they were originally used by the farmer.

6 What damage could be done by water flowing out from old lead mines?

7 Burning rubbish so that we do not have to bury it is not always a good idea. Why?

Summary

✱ Food from the Sun

» Sunlight provides energy for all living things.

» Photosynthesis uses light energy to make sugar and oxygen from carbon dioxide and water.

» Plants are called producers because they produce food. Animals eat plants to get this food. They are called consumers.

» Animals that eat plants are called herbivores. Animals that eat other animals are called carnivores. Food chains show how energy passes from plants to herbivores and on to carnivores. The food chains in an area link into a food web.

» Plants often compete with each other for light and space, and for water and nutrients from the soil.

✱ Recycling wastes

» All materials in the living world are recycled to provide the building blocks for future organisms.

» Decay of dead plants and animals by microorganisms returns carbon to the atmosphere as carbon dioxide to be used by plants in photosynthesis. Other nutrients are released to the soil to help plants grow.

✱ Keeping the planet safe

» Human activities can cause pollution of rivers and oceans through sewage, fertiliser or toxic chemicals.

» Air pollution includes problems caused by smoke and from gases such as sulfur dioxide, which contributes to acid rain.

» Pollution on land comes from landfill (burying of rubbish) and from poisonous chemicals such as pesticides and herbicides, which may be washed from the land into streams, ponds, rivers and oceans.

Show you can...

A Have a discussion with a partner about recycling. Why is it a good idea? What do you recycle? Agree two ways you could improve the number of things you recycle.

B Produce a poster to show a design for a greenhouse to grow plants like cacti that need a hot, dry environment. Your greenhouse should be suitable for someone living in the north of England. Include plenty of labels to explain how the greenhouse features help the plants to grow well.

C Choose an area of waste ground in your neighbourhood. How could you convert this into a nature reserve? Take photographs and create diagrams and drawings to show what is possible. Create a presentation of your ideas to convince the local council to fund the project.

D How could you find out how the level of light affects how much oxygen is produced by algae? Prepare a paper to explain how you would do it and suggest what results you would expect, with reasons for your prediction.

E Build and test a simple machine to water houseplants. It should be possible to set the machine to water the plants every 12 hours.

▲ **This is one way to water the garden plants!**

Episode 4 Sex and survival

✱ Animal attraction

▲ **Beauty contests: why do we have them?**

There are lots of beauty contests in the animal world. A robin will sing as loudly as possible to attract a mate. Walruses will fight to the death to protect their females. This is to ensure that they produce offspring.

▲ **It's all about reproduction for robins and walruses!**

Reproduction

There are two sorts of reproduction: asexual reproduction and sexual reproduction.

Table 1 Asexual and sexual reproduction

	Asexual reproduction	Sexual reproduction
What reproduces in this way?	Many plants and microorganisms reproduce asexually.	Animals usually reproduce sexually. Many plants can reproduce sexually as well.
How many parents are needed?	One parent is often enough.	Two parents are usually needed: male and female.
Are the offspring the same as the parents?	The offspring are the same as the parents. They are clones of the parent.	The offspring are slightly different. Some are better and some not so good.
What are the advantages?	A successful parent can produce lots of identical offspring at one time.	Some offspring might fit better in the new environment.
What are the disadvantages?	If the environment changes, the offspring may not fit in the new conditions.	Good parents can produce weaker offspring.

▼ Asexual reproduction

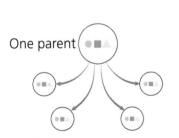

One parent

All offspring are the same.

▼ Sexual reproduction

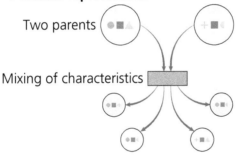

Two parents

Mixing of characteristics

Offspring are often different to each other and the parents.

Sexual differences

Sexual reproduction needs special cells called gametes. Gametes are cells that can join together in sexual reproduction.

The male produces gametes called sperm. The female gametes are the eggs. Gametes are produced by organs called gonads.

In human beings the male gonads are the testicles. They hang just outside the body in the scrotum. Females produce eggs. In humans these are made in the ovaries that are buried deep in the body.

1 Will a very beautiful man or woman always produce beautiful children?

2 What does the word reproduction mean?

3 a Give one advantage of sexual reproduction.

 b Give one advantage of asexual reproduction.

4 It is often difficult to tell the difference between male and female babies. It is easier to tell the difference as they grow older. Why is this?

5 Give two ways the sperm is adapted to get to the female.

6 Give one adaptation of the egg to support the growing baby.

Chromosomes

It takes 9 months to go from a fertilised egg to a new baby. The fertilised egg contains all of the instructions that show how to make a baby.

The instructions are coded in chemicals called genes. Each gene tells the body how to make an important chemical.

Genes are organised into chromosomes, a bit like a book organises pages.

Humans have 23 pairs of chromosomes. That is like having two copies of 23 books.

Chromosomes look like tiny worms. Most chromosomes contain a collection of different genes that are not related. However, two chromosomes contain many of the genes that affect many of the male–female differences. These are called the sex chromosomes.

In males, the sex chromosomes consist of one long chromosome called the X chromosome and a shorter one called the Y chromosome.

In females there is no short Y chromosome. There are just two long X chromosomes.

Gametes only contain one from each pair of chromosomes. So, all eggs contain one X chromosome. Half of the sperm contain an X chromosome and half contain a Y chromosome.

▲ **A single fertilised egg contains about 25 000 genes. That's more instructions than the most complicated Lego model ever!**

7 How many chromosomes in total are there in a normal human cell?

8 How many chromosomes does a single sperm cell contain?

▼ **XX and XY chromosomes can produce X and X eggs and X and Y sperm**

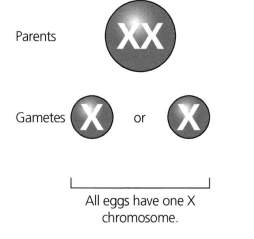
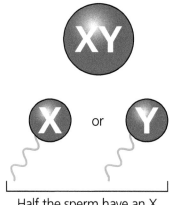

Parents

Gametes

All eggs have one X chromosome.

Half the sperm have an X chromosome and half have a Y chromosome.

Episode 4 Sex and survival

46

What are genes made of?

Genes are small lengths of a chemical called DNA. This is a very complicated chemical that is able to code instructions for the cell.

You can think of genes as a recipe for making a cell. Think about a recipe book telling you how to make an omelette, or a chocolate tart. If the genes are the recipes, then the DNA is the language the recipes are written in.

The really clever thing about DNA language is that it only has four letters! Imagine a recipe book for cooks that could only use four ingredients!

Genetic engineers

Cystic fibrosis is an illness caused by a damaged length of DNA. A person with cystic fibrosis has a damaged CTFR gene. They have difficulty breathing and cannot digest some foods properly. In the past, they often did not live beyond 20 years of age.

Nowadays, with good treatment, they can live an almost normal life. There are 10 000 people in the UK who have this disease. Genetic engineers are looking for a way to cut out the CTFR gene from a healthy person and give it to someone with cystic fibrosis. This could cure them of the disease.

▲ **Instructions to make this fetus grow from a fertilised egg are coded in genes**

9 What are some of the benefits of genetic engineering?

10 Some people don't agree with genetic engineering. Why do you think this might be?

▼ **The stages in genetic engineering**

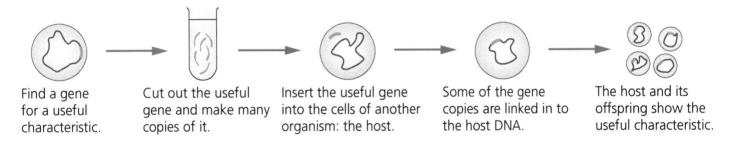

Find a gene for a useful characteristic.

Cut out the useful gene and make many copies of it.

Insert the useful gene into the cells of another organism: the host.

Some of the gene copies are linked in to the host DNA.

The host and its offspring show the useful characteristic.

Pets and plants

✳ A suitable pet

Emily has a pet royal python, called Mike. Pythons like this come from the hotter areas of Africa. Mike is used to high temperatures and dry conditions.

Emily lives in Bolton where it is often cool and wet. She needs to make sure that Mike has the conditions he needs. So, Mike lives in a giant heated glass case. He will grow to about 1.5 metres long and eat a live mouse once a week.

1 What pet would you choose? Why?

2 Would it be adapted to the area where you live?

▲ **Mike has the conditions he needs to survive in a heated glass case**

Adaptations

A royal python is at home in the forests and grasslands of central Africa. He fits very well into that environment. Scientists say that he is adapted to his environment. All living things have adaptations for their environment. Any living thing that does not have good adaptations will tend to die out.

Royal python

Where do they live?	Central Africa – hot and dry areas
What do they eat?	Small rodents that live in the grassy areas of central Africa
How do they eat?	Hunting
Adaptations?	Can survive in dry conditions Eats only once a week

Polar bear

Where do they live?	Arctic conditions – cold and covered with ice
What do they eat?	Fish and seals from the waters of the Arctic Ocean
How do they eat?	Hunting
Adaptations?	Thick fur protects against cold. White fur camouflages it against the snow. Sharp claws and teeth can kill animals

Camel

Where do they live?	North Africa – hot and dry deserts with low rainfall and few plants
What do they eat?	Plants that manage to grow in the areas of the desert that have water
How do they eat?	Grazing
Adaptations?	Fur protects against bright sun and overheating Hump can store fat for times when it is not able to find food Feet are very wide to make it easier to walk on the sand

Tiger

Where do they live?	India and Burma
What do they eat?	Other animals, including people
How do they eat?	Hunting
Adaptations?	Its skin provides camouflage in the jungle Sharp claws and teeth can kill animals

3 Give two ways the royal python is adapted to life in the African forests.

4 Give two ways a polar bear is adapted to life in the Arctic.

5 Explain why a tiger would not survive well in the Arctic.

Plants and adaptations

Plants have adaptations too! Some grow better in full sunlight, others need more shade. Some survive dry periods but others wilt if you do not water them for a day. Gardeners need to know what plants work where.

Bluebells

Where do they live?	In damp forests; the flowers come out before the leaves on the trees above them
How much water do they need?	The plants cope with wet conditions; they are easily damaged by dry periods
How much sunlight do they need?	Shade
How much cold can they survive?	They survive some frost by spending the winter as bulbs under the ground

Cacti

Where do they live?	Drier areas, particularly deserts
How much water do they need?	Almost none, watering can actually damage them
How much sunlight do they need?	Full sun
How much cold can they survive?	Almost none; they are easily damaged by frost

6 a Give two ways that cacti are adapted to life in very dry regions.

b What would a gardener need to grow the best cacti?

7 Christmas trees are types of fir tree.

a What conditions would be good for a Christmas tree farm?

b East Anglia has less rainfall than most areas of the UK. Would this be a good place to start your Christmas tree farm? Why?

Fir trees

Where do they live?	In forests in the northern areas of the UK
How much water do they need?	They need a lot of water; the leaves are waxy to reduce water loss but can go brown and die in dry conditions
How much sunlight do they need?	Full sun or partial shade
How much cold can they survive?	Can survive frost over a long period of time

✳ Life story

Begin at the beginning

Life probably started in a warm ocean billions of years ago. We do not know how this happened. Strong ultraviolet light that reached the Earth's surface then might have been important.

The first simple living organisms produced offspring that were slightly more complicated. Over billions of years living things became more and more complicated. Eventually there were many different sorts of living things. Scientists call this evolution.

▲ Now they are plastic toys for children, but once these monsters ruled the planet!

Breeding like rabbits

A pair of animals can produce many more offspring than they need to replace themselves. A pair of rabbits who meet up on January 1st could be a family of nearly 100 children and grandchildren by Christmas!

Not all of these offspring will survive. Some will be sick. Others will be eaten by predators. Some will fail to find food. Even the ones that become adults may not produce offspring of their own. There are so many ways to fail.

8 Blue tits produce 8 to 10 eggs every year. By the following year, only 2 or 3 birds have survived. Give three reasons why the blue tits might have died.

Not all the same

Rabbits, like people, are not all the same. There are small differences between them. Biologists call these small differences variation.

▲ **This Victorian family had 14 members! They were related, but they were all slightly different.**

These slight differences are due to sexual reproduction. Everyone inherits a slightly different mix of genes from their parents. But sometimes a completely new gene is created. This is called a mutation. Mutations make the offspring slightly different.

Natural selection

Imagine some rabbits had slightly thicker fur. They could cope better in the cold. If there were many bad winters, all the rabbits with thin fur would die out and the thicker-furred ones would take over. Scientists call this the survival of the fittest or natural selection.

Imagine there were other changes in the rabbits as well. Over many thousands of years, some rabbits could change so much that they don't even look like rabbits! They have become a new type of animal altogether. This is an example of evolution.

9 a Even members of the same family are not exactly the same. Give two reasons why.

  b What is a mutation?

Wolf or dog?

About 40 000 years ago, most of Europe was covered with ice. Times were hard. Wolf cubs were taken from their parents and reared as hunting wolves. They might have helped to hunt woolly mammoths. They may also have been fed scraps and kept as pets. In the harsh winter when food was hard to find, a nice, fat pet wolf made a good stew as well!

Then humans started to decide which of their pet wolves to breed. They might have picked the fastest wolf. They might have picked the wolf with best hearing. Maybe they just picked a wolf that they thought looked good.

Over thousands of years the original wolves became dogs. Today, there are many different types of dog, but they all come originally from the grey wolf. These dog breeds were created by humans. This is not natural selection. It is artificial selection.

▲ **Wolves were used to help humans to hunt in the Ice Age**

Table 1 Animals and their ancestors

	Pigs	Dogs	Sheep
What did the first animal look like?	Wild boars were large, very strong animals with wiry bristles and a hairy coat. Their tusks could be 10 cm in length!	The ancestors of dogs were wolves. These were wild animals that lived in packs.	The original wild sheep tended to have short tails and long legs. The colour of the wool was always a sort of light brown.
What does the modern animal look like?	Modern pigs are almost hairless, the tusks are much smaller and many have lots of fat under their skin.	Modern dogs have many different breeds, ranging from small Poodles with curly hair to large Alsatians with straight fur.	Domesticated sheep tend to have shorter legs and longer tails. This means that they are less able to jump out of the field!
What have we selected for in breeding?	Quiet characters. Good taste!	Different breeds of dogs have been selected for hunting, herding sheep or just as pets for the family.	Different colours in the fleece – particularly white.

Summary

✱ Reproduction

» Reproduction is the production of offspring to increase the local population.

» Sexual reproduction needs male and female sex cells to join. There is a mixing of genetic information, which leads to variety in the offspring.

» Asexual reproduction needs only one parent. There is no mixing of genetic information, which leads to identical offspring called clones.

» The sperm and egg are adapted to make sure they meet to produce the fertilised egg.

» Organisms often display special behaviours to find and keep a mate. Robin red breasts sing very loudly, male walruses can fight with other males.

✱ Genetics

» Genes control how an organism develops from a single fertilised egg. They are the instructions that show how to create all the parts of the organism.

» Genes are made of a chemical called DNA, which is contained in structures called chromosomes. Chromosomes carry genes that control the characteristics of the body.

» Humans have 23 pairs of chromosomes. One pair carries the genes that determine sex: females have the same sex chromosomes (XX); in males the chromosomes are different (XY).

» Genetic engineers 'cut out' useful genes from some chromosomes and transfer them to the cells of other organisms.

✳Adaptations and evolution

» Animals and plants are adapted to survive where they normally live. If they are moved to other areas, they may die out.

» Changes to the environments of animals and plants can be caused by non-living factors (e.g. a change in temperature) or by living factors (e.g. new predators move into the area).

» In natural selection, individuals with characteristics most suited to their environment are most likely to survive to breed successfully.

» In artificial selection (selective breeding) humans choose which plants and animals produce offspring. These are selected for usefulness or appearance.

» The theory of evolution says that all living things evolved from simple life-forms by mutation and natural selection.

» Life probably started in the oceans more than three billion years ago.

Show you can...

A Explain to a partner the differences between sexual and asexual reproduction.

B Produce a poster to show four different breeds of dog. Label the important differences between them and explain why humans have selected these features, for example a husky has a very thick coat to keep it warm in sub-zero temperatures.

C Genetic engineering is a controversial issue. Prepare a 2-minute radio news item to explain how scientists are hoping to cure cystic fibrosis using gene therapy. Your listeners should hear the words 'gene', 'chromosome' and 'DNA' in your broadcast.

D Rabbits that live in the Arctic have much thicker fur than rabbits that live in England. Why do you think this is? How could you test your idea?

E Build a model of an enclosure for a zoo. Decide which animal it will contain and then create a model to show what the enclosure would look like. Explain how the animal's adaptations would help it to fit into your enclosure. Remember, you want visitors to be able to see the animals without disturbing them too much.

Episode 1 Substances

The pure stuff

To a chemist, pure means something that contains only one substance. So, pure water contains nothing but water. Pure gold is nothing but gold.

A pure substance has a fixed composition and properties. A fixed composition means that it always contains the same things. Fixed properties means that it always behaves in the same way.

▲ **Do you think this gold is pure?**

Different mixtures

A cup of coffee is never pure! It contains many chemicals that give it a coffee flavour, and boiling water. Some people add milk and sugar to make it even more complicated.

A mixture contains a number of different substances. Mixtures can have different amounts of each substance. The substances in a mixture are not chemically bonded.

Chocolate cake is a mixture. It contains flour, sugar, fat and chocolate.

1 Write a sentence with the word 'substance' in it.

2 What do chemists mean when they say a substance is pure?

▲ **Your friend's espresso may be strong and have no sugar or milk. Your cappuccino might be sweet and milky, but they are both coffee**

▼ *Orange Life* and *Orange Sparkle* – but are they the same?

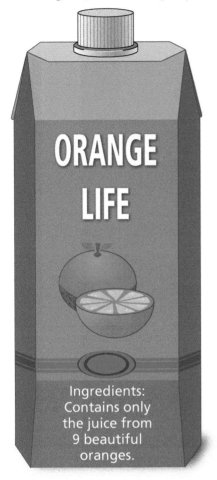

ORANGE LIFE

Ingredients:
Contains only
the juice from
9 beautiful
oranges.

ORANGE SPARKLE

Ingredients:
Orange juice,
artificial sweetener,
colour, preservative,
carbon dioxide, water.

3 Write a sentence that contains the word 'mixture'.

4 Give one difference between a mixture and a substance.

5 Are these two types of orange juice the same substance?

Separating mixtures

Chemists often have to separate mixtures. Some things they may have to separate are:

✷ An insoluble solid from a liquid

✷ A soluble solid from a liquid

✷ A liquid from a mixture of liquids

✷ Pure sugar from sugar beet by crystallisation.

An insoluble solid from a liquid

Tiny grains of gold are washed out of the rocks by streams. They cannot dissolve in water and so settle out. If you are lucky you can find them and make your fortune.

A soluble solid from a liquid

These ponds are filled with seawater. As the water evaporates the salt is left behind. You can collect the salt to sell.

A liquid from a mixture of liquids

In the 1930s, all alcoholic liquor was banned in the USA. But people used to make whisky with illegal stills like this one. The fermented mixture is poured into the bottom and heated. The vapours that come off are then cooled to make a liquid. That is the whisky. They used to call it moonshine!

Pure sugar from sugar beet by crystallisation

These workers is the 19th Century are making sugar crystals. To do this, boil up sugar beet in water to dissolve the sugar. Filter out the mashed beet and then boil the clear solution to make the sugar crystallise out.

Chromatography

Chromatography separates out dissolved substances in a solution. It can be used to check the colour of inks.

How does paper chromatography work?

▼ **The process of paper chromatography**

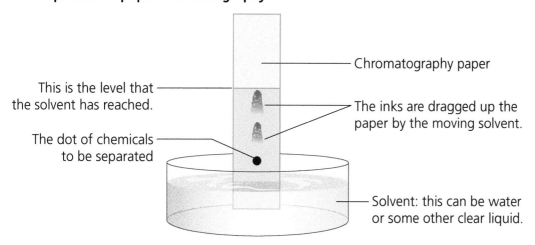

Chromatography paper

This is the level that the solvent has reached.

The inks are dragged up the paper by the moving solvent.

The dot of chemicals to be separated

Solvent: this can be water or some other clear liquid.

Practical

The perfect colour?
Many coloured inks are mixtures. To get that perfect red colour for example, you may need two or three types of red. Use paper chromatography to find out if some of the inks you use are pure substances or mixtures.

Activity

Safe to drink?
All over the world, people have to get their drinking water from streams and rivers. Often the streams and rivers have mud and gravel mixed in and bits of leaves and rubbish floating in them. Build a water filter to separate out clear water from dirty water. Would your clear water be safe to drink? Why? Why not?

6 Which separation technique could you use to separate:

a the sugar from a cup of coffee

b the coffee grounds from the coffee liquid

c the mix of colours used in a pot of ink?

Episode 2 Melting and evaporating

✴ Liquid chocolate

Do you like thick chocolate on cakes? Really hot chocolate is poured over the cake. As the chocolate cools it solidifies.

1 The chocolate on a chocolate cake has to be hot when it is poured over the filling. Why?

2 What might happen if the melted chocolate was too hot?

3 What happens to the chocolate that drips off the cake?

4 Creme eggs have a soft centre. How do you think they make them?

▲ This chocolate egg has a soft filling and a hard outside coating of chocolate

Solids, liquids and gases

Substances around us can exist in three states: solid, liquid or gas. They all contain the same substance but they look different.

The temperature at which a solid becomes a liquid is called its melting point. The temperature when the liquid changes to a gas is called the boiling point.

Table 1 The three states of matter

Solids	Liquids	Gases
Solids do not change shape and have a fixed volume.	Liquids change shape to fit the bottom of the container and can flow.	Gases change shape to fill the container, can flow and do not have a fixed volume.

Explaining melting

Melting is a change of state. The chemicals in the solid chocolate and the melted chocolate are the same. So, why is one a liquid but the other is a solid?

▼ **Differences between solids and liquids**

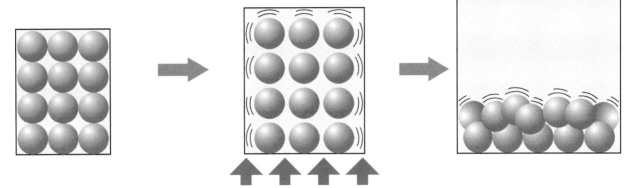

Heat

Particles in a solid are close together. They are stuck in a regular pattern. Strong forces hold them in place. They cannot move around.

To melt a solid, heat it up. The energy in the heat makes the particles vibrate faster. The particles begin to move around each other. The solid becomes a liquid – it melts.

The particles of a liquid are not in a regular arrangement. They move around each other. This is why a liquid can flow. There is very little space between the particles so it is difficult to compress a liquid.

5 Give two differences between a solid and a gas.

6 Name three substances you know that are liquids at room temperature.

7 Name three substances you know that are gases at room temperature.

8 Draw two labelled diagrams to show the particles in ice and liquid water.

Practical

Investigating melting
What temperature does chocolate melt at? Plan and carry out an experiment to find out. How would this be useful to the makers of chocolate fountains?

✱ Stinkers!

Smell like Elvis

You can buy anything online, including Elvis's sweat! Only £7! But why would you want to smell like Elvis? And what causes the smell?

9 Do you think samples of liquid that are for sale are really Elvis's sweat? Give reasons for your answer.

10 You should never smell an unknown chemical in a science laboratory. Why?

▲ What causes Elvis's sweat to smell?

Practical

Investigating evaporating
Find out how temperature affects how quickly a liquid evaporates. How could you use this information to work out how long your sample of Elvis's sweat would last?

Distillation

Jebel Ali Power and Water Plant boils sea water and condenses the steam to make fresh water. It produces 140 million gallons of fresh water every day. This provides enough for drinking, cooking and washing. There is often enough left over to water parks and golf courses and even dump in the sea!

11 Write a sentence using the word 'evaporation'.

▲ Jebel Ali Power and Water Plant

Explaining evaporation

When a liquid evaporates it does not change the substances it contains. The particles simply break away from each other. The heat makes this happen. If the heat is removed the gas can condense again to make a liquid

▼ **How evaporation happens in a liquid**

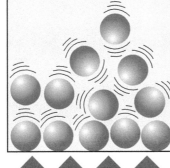

Heat

In a liquid the particles can move around but are tightly packed together. So you cannot compress a liquid.

When you add heat the particles move around more quickly. The heat gives the particles energy to move around more quickly.

To evaporate a liquid add more heat. Some particles escape the liquid altogether. They move around freely. They have become a gas. There is so much space between particles that a gas can be compressed.

Better smells

Perfumes are mixtures of chemicals. They evaporate from the body surface. When these chemicals enter our nose we detect them as a nice smell. Some are extracted from rose petals or other flowers. Some are made in factories or even oil refineries.

▲ Musk, a powerful natural scent, used to be collected from glands near the testicles of wild musk deer to make perfume!

12 Why does steam take up so much more space than liquid water?

13 What is your favourite smell? Why?

14 Natural gas has no smell at all. Why do natural gas suppliers add a smell, and a really strong bad one?

Practical

Orange peel perfume
Steam distillation is a good way to extract the scent from orange peel or lavender flowers. Investigate how well the method works. See how strong you can make your extract.

The first artists?

▲ **The oldest piece of art ever discovered is in a cave in South Sulawesi. The prints of hands shown are probably 40 000 years old!**

The first magicians?

Why did our ancestors paint cave walls? Maybe it showed they owned the cave? Maybe it was a sort of magic? Maybe they just got bored with plain walls?

◀ **Was this done by a chemist?**

The first chemists?

Some paint is washed off by water. Some paints seem to stick to the cave wall. Chemists use the word react to describe an important change in substances. When two or more substances react together they make new substances.

The paint and the wall have reacted together to make a new compound. This is not washed off so easily. So, these ancestors were the first artists *and* the first chemists.

?

1 How do you think these paintings were made?

2 Why do you think our ancestors painted the walls where they lived?

3 How is this similar, or different to, graffiti artists nowadays?

Practical

Making your mark
Test different powders and ground up rocks to see which leaves the best mark on a cave wall. Would your artwork last 40 000 years?

?

4 Write a sentence with the word 'react' in it.

5 Give two other changes that you know that could be chemical reactions.

From alchemy to chemistry

Alchemists lived hundreds of years ago. Many were like magicians with special secrets and magic spells. They searched for a medicine that would let them live forever or turn lead into gold. The alchemists failed, but laid the groundwork for modern chemistry. Even the word 'chemistry' comes from the word 'alchemist'.

Modern chemists depend on experiments and observations, not on secret rules or magic spells. They are making new plastics, drugs to treat disease or chemicals to kill pests in your garden. Some may even be testing paint on walls!

Atoms

Modern chemists believe that everything is made up of tiny particles called atoms. These are so small that the dot of ink you can see at the end of this sentence contains over 20 billion atoms. This is almost enough for everyone on the Earth to have three each!

Elements and compounds

There are just over 100 different types of atoms. If a substance contains only one type of atom it is called an element. Since there are only 100 types of atom, there are only about 100 elements. Most substances are compounds. These are made of two or more types of atoms linked together by chemical bonds. When atoms link together they react, like the reaction between the paint and the cave wall.

6 What do you think chemists should be working on today? Why?

7 Write a sentence with the words 'atom' and 'element' in it.

8 Which of the substances in the diagram is an element?

The water contains two elements joined together: hydrogen and oxygen.

The metal in the tank contains a single type of atom: iron.

The reef contains mainly calcium carbonate. This contains three types of atoms joined together: calcium, carbon and oxygen.

▲ Water is a compound that is made of two elements linked together: hydrogen and oxygen

Secret names

We use names to tell people who or what we are talking about. What if you want to keep some secrets? That's when you use a code name. Each of the people or events below had code names. Can you work out which code name fits each one?

Barack Obama	Queen Elizabeth II	James Bond

The project to make the first atomic bomb	Nintendo Wii	The first iPhone
		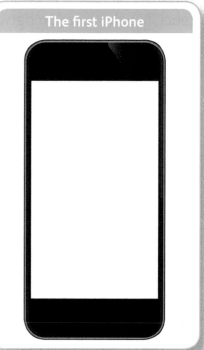

Here are the code names:

007
Dolphin
Kittyhawk
Manhattan Project
Purple 2
Renegade

9 a Match the code names to
      the correct people or events.

   b Why might people use a code
      name and not the real name?

   c Think of a great code name
      for yourself. Why have you
      chosen that one?

Chemical names

Chemists use code names for substances. But the code is not
secret or difficult to learn. The code lists what is in the compound.

So, what substances are in sodium chloride?

* The first word 'sodium' tells us that it contains the metal
  sodium.

* The second word 'chloride' is more complicated. It tells us
  the compound contains chlorine.

Table 1 explains the most common 'second words' you might
see in chemical names.

Table 1 Working out chemical names

If it is a …	It contains…	For example
chloride	Chlorine	Sodium chloride
nitrate	Nitrogen and oxygen	Potassium nitrate
sulfate	Sulfur and oxygen	Copper sulfate
carbonate	Carbon and oxygen	Calcium carbonate
oxide	Oxygen	Iron oxide
hydroxide	Hydrogen and oxygen	Sodium hydroxide

Chemical equations

When two substances react together the atoms are
rearranged and new substances form. Chemists use
word equations to show this change:

sodium + chlorine → sodium chloride

carbon + oxygen → carbon dioxide

potassium + chlorine → potassium chloride

When you read chemical equations you can see what
chemicals you start with and what you end up with. The
metal never changes its name. The other part may change.

What would this equation finish with:

zinc + oxygen → zinc…

10 Why do chemists need
   clear names for the
   substances they use?

11 a Sand on the beach
      is made of silicon
      dioxide. What two
      elements are present
      in silicon dioxide?

    b If you see 'ate' at the
      end of a chemical
      name, what element
      do you think it
      contains?

Summary

✳ Substances and mixtures

» A pure substance has a fixed composition and always behaves in the same way (fixed properties).

» A mixture contains two or more substances that are not chemically combined together. A mixture can include different substances in different amounts at different times. Two mixtures will behave slightly differently.

» Chemists can separate mixtures by filtration, distillation, crystallisation and chromatography.

» Paper chromatography can be used to separate mixtures and can give information to help identify substances. In paper chromatography, water or another liquid moves through the paper carrying different compounds different distances.

✳ Solids, liquids and gases

» Solids have a fixed shape and volume. Liquids can change their shape but keep the same volume. Gases change shape and volume.

» A solid is converted to a liquid by heating. We call this change melting. Heating a liquid converts it to a gas. We call this change evaporation.

» The particles in a solid are held rigidly in place. The particles in a liquid are close together but can move around each other like people in a crowd. The particles in gases are separated by lots of empty space and can move around freely.

Chemical reactions

» All substances are made up of very small things called atoms. There are about **100** different types of atoms.

» A substance that contains only one type of atom is called an element. There are about **100** elements.

» Most of the things we see around us are compounds. Compounds are made when different sorts of atoms join together. This is called a chemical reaction. Water is a compound made when two hydrogen atoms react with one oxygen atom and join together.

» Chemicals are named after the elements they contain.

» Chemical equations show what happens in a reaction in words or symbols.

Show you can...

A Explain to a partner how to crystallise salt from seawater. Use these clue words in your discussion: boil, crystallise, dissolve, solvent, solute, solution.

B Design a poster to show how to produce a supply of clean water from a sample of dirty water.

C Create an animation to show what happens to the particles when water boils and steam condenses.

D Prepare a paper showing how to accurately find out the melting point of wax for a candle sample.

E Design and build the water supply part of a survival kit to keep people alive in disaster situations. It should produce clean, drinkable water from dirty water. It must be cheap, lightweight and easy to use.

Episode 4 Sorted!

✱ Metal or non-metal?

All elements are either metals or non-metals.

Which of these elements do you think are metals? Which are non-metals?

Copper

Magnesium

Carbon

Sulfur

Iron

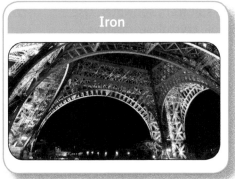

Oxygen

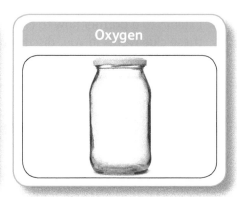

The metals

Most of the elements are metals. Metals are:

✱ solid   ✱ strong   ✱ shiny.

Metals conduct electricity and heat well and can be beaten into sheets or pulled into wires. They have many uses.

Think about the aluminium on your smartphone, the iron in nails and screws, or the copper in electrical wires.

▲ Gold can be beaten into very thin foil called gold leaf

The non-metals

Non-metals are very different to metals and to each other. They are not all the same colour. Chlorine and oxygen are gases. Carbon and iodine are solids. Bromine is a liquid at room temperature. Non-metals:

* include gases, liquids and solids

* do not conduct heat or electricity well

* cannot be made into wire or beaten into flat sheets.

Non-metals are important. Most of your body is made up of compounds of non-metals.

▲ This is one of the world's biggest diamond, the Koh-i-noor. It was owned by an Indian prince, who put it in a temple as the eye of a statue of a Hindu goddess. It was taken, and is now part of the Crown of Queen Elizabeth The Queen Mother, which is used for coronations in the United Kingdom. It is a lump of the non-metal carbon

▲ Neon is a gas used in lights like these. It does not react with anything else.

▲ Chlorine was the first ever chemical weapon. It was used in the First World War and killed over 11 000 soldiers and injured many more. Nowadays, a solution of chlorine in water can kill dangerous microbes, and has probably saved millions of lives across the world. You will have smelt chlorine in bleach and swimming pools

Activity

Describing elements
Make observations about the selection of elements you are given. How will you recognise the metals and non-metals?

Get into groups!

▼ The chip shop organises what it can sell into groups. This makes it easy for customers to see what they want

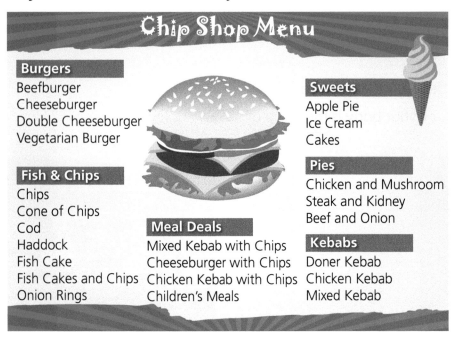

Chip Shop Menu

Burgers
Beefburger
Cheeseburger
Double Cheeseburger
Vegetarian Burger

Fish & Chips
Chips
Cone of Chips
Cod
Haddock
Fish Cake
Fish Cakes and Chips
Onion Rings

Meal Deals
Mixed Kebab with Chips
Cheeseburger with Chips
Chicken Kebab with Chips
Children's Meals

Sweets
Apple Pie
Ice Cream
Cakes

Pies
Chicken and Mushroom
Steak and Kidney
Beef and Onion

Kebabs
Doner Kebab
Chicken Kebab
Mixed Kebab

?

1 How could you sort these things into groups:

a the games in a video games shop

b magazines

c a collection of plastic figures?

The periodic table

The periodic table organises all the elements into groups. Elements with similar properties are in the same column. The columns are called groups. You can see the periodic table on the next page.

Group 1: the alkali metals

Group 1 metals do not look like aluminium or copper. They are quite soft, a bit like modelling clay. Group 1 metals react with the oxygen in the air around them very easily. They burn brightly to make a compound that makes a strong alkali in water. The further down the group you go, the more reactive they are.

Group 7: the halogens

Group 7 contains the halogens. Chlorine is in group 7. You will have smelt it in bottles of bleach or at the swimming pool. Chlorine helps to kill microorganisms that can cause disease. Iodine is another element in group 7. A solution of iodine in alcohol is often washed over the skin before a surgeon makes a cut.

▲ Potassium, from Group 1, bursts into a purple flame when you drop it in warm water

▲ The brown colour shows where the iodine solution has been used to sterilise the skin. Iodine is found in Group 7.

Episode 4 Sorted!

Group 1	Group 2											Group 3	Group 4	Group 5	Group 6	Group 7	Group 0
H hydrogen																	**He** helium
Li lithium	**Be** beryllium											**B** boron	**C** carbon	**N** nitrogen	**O** oxygen	**F** fluorine	**Ne** neon
Na sodium	**Mg** magnesium											**Al** aluminium	**Si** silicon	**P** phosphorus	**S** sulfur	**Cl** chlorine	**Ar** argon
K potassium	**Ca** calcium	**Sc** scandium	**Ti** titanium	**V** vanadium	**Cr** chromium	**Mn** manganese	**Fe** iron	**Co** cobalt	**Ni** nickel	**Cu** copper	**Zn** zinc	**Ga** gallium	**Ge** germanium	**As** arsenic	**Se** selenium	**Br** bromine	**Kr** krypton
Rb rubidium	**Sr** strontium	**Y** yttrium	**Zr** zirconium	**Nb** niobium	**Mo** molybdenum	**Tc** technetium	**Ru** ruthenium	**Rh** rhodium	**Pd** palladium	**Ag** silver	**Cd** cadmium	**In** indium	**Sn** tin	**Sb** antimony	**Te** tellurium	**I** Iodine	**Xe** xenon
Cs caesium	**Ba** barium	**La*** lanthanum	**Hf** hafnium	**Ta** tantalum	**W** tungsten	**Re** rhenium	**Os** osmium	**Ir** iridium	**Pt** platinum	**Au** gold	**Hg** mercury	**Ti** thallium	**Pb** lead	**Bi** bismuth	**Po** polonium	**At** astatine	**Rn** randon
Fr francium	**Ra** radium	**Ac**** actinium	**Rf** rutherfordium	**Db** dubnium	**Sg** seaborgium	**Bh** bohrium	**Hs** hassium	**Mt** meitnerium	**Ds** darmstadtium	**Rg** roentgenium	**Cn** copernicium	**Uut** ununtrium	**Fl** flerorium	**Uup** ununpentium	**Lv** livermorium	**Uus** ununseptium	**Uuo** ununoctium

*	**Ce** cerium	**Pr** praseodymium	**Nd** neodymium	**Pm** promethium	**Sm** samarium	**Eu** europium	**Gd** gadolinium	**Tb** terbium	**Dy** dysprosium	**Ho** holmium	**Er** erbium	**Tm** thulium	**Yb** ytterbium	**Lu** lutetium
**	**Th** thorium	**Pa** protactinium	**U** uranium	**Np** neptunium	**Pu** plutonium	**Am** americium	**Cm** curium	**Bk** berkelium	**Cf** californium	**Es** einsteinium	**Fm** fermium	**Md** mendelevium	**No** nobelium	**Lr** lawrencium

Key
- ☐ Metal
- ▨ Non-metal
- ▨ Difficult to classify

▲ **The periodic table of elements**

Episode 5 Metal treasures

✱ Ancient metals

A tale of three finds

The metal objects shown here were found by metal detectorists.

▲ A torc is a thick twist of gold worn around the neck. The original owner of this torc was probably a Celtic chief living in Norfolk over 2000 years ago

▲ A metal coin from around 800 years ago. The metal is badly corroded and stained brown

▲ An ancient Chinese coin from almost 3000 years ago. The corrosion is clear to see.

Corrosion

All metals react with oxygen to make oxides. When iron reacts with oxygen and water over a period of a few weeks, it makes a brown substance that we call rust. Slowly the metal corrodes. The corrosion we see on a metal is often a mixture of chemicals. Gold is unusual. It is very unreactive so can remain in the ground for many years and not change.

Practical

Investigating corrosion
What do iron nails need to rust? Carry out an investigation to find out what speeds up or slows down rusting.

?

1 What is the most valuable thing that you have ever found?

2 What do we call the brown staining on the two metal coins?

3 Why does the golden torc look so good, even though it is much older than the horseshoe?

Extracting metals

You can find gold as a pure metal in nature. Almost all other metals are found in compounds. The metal-containing compound is called a mineral. It is usually mixed in with other rocks as well. The mix of useful mineral and other rock is called an ore. Galena is an ore that contains a compound of lead and sulfur. Getting lead out of galena is not easy!

4 Which of the processes in the extraction of lead from galena are chemical reactions?

Rocks and ore are dug out of ground.

Ore is crushed to a powder in giant machines.

Mineral is extracted. Galena floats in the froth on top of a special liquid but the other chemicals sink to the bottom.

Powdered ore is heated in a blast furnace. Oxygen is blown through the hot mixture.

The oxygen reacts with the galena to make lead oxide and give off sulphur dioxide.

The air supply is closed off. The lead oxide breaks down in the heat to give lead metal.

Molten lead cools and solidifies to make blocks of lead.

Recycling metal

Finding, extracting and purifying ore to produce metal ingots costs a lot of money and energy. It is much better to collect old metal items and melt down the metal to use it again. If you recycle your old baked bean tin or drinks can it may come back to you as a car or a washing machine.

5 Give three advantages of recycling metals instead of always producing new metal from ores.

Problems and protests

Metals are useful chemicals. We use them for houses, cars, jewellery and even in some medicines. However, digging metals out of the ground can cause environmental problems. Is it ok for some people to put up with the problems caused by mining and metal extraction if other people many miles away enjoy the benefits of the metal?

> The ore here is quite low quality. When they separate out the waste they pile it up in slag heaps. When rain seeps through these heaps, it dissolves the poisonous compounds in them. These get into the rivers and poison the fish.

> The heavy lorries are destroying the roads around here! And they are dangerous to children and old people crossing the road!

> This used to be a beautiful green valley. Now look at it! It is just a waste of spoil heaps.

> Global Mining tries to reduce any damage as much as possible. We plant trees to screen our waste heaps. Our lorries do not work at night. We have also given money for a new library for the primary school. We pay good wages to many local people. The metals we find here are essential to our country's prosperity. Many of the things these protestors depend on every day would not exist if Global Mining could not extract the metals needed to build them.

6 Give three possible environmental problems caused by mining.

7 What are the advantages to an area if metal ore is discovered there?

Why are metals so useful?

Metals:

* are strong

* are shiny

* conduct electricity and heat well

* have high boiling points

* can be beaten into sheets or pulled into wires.

These properties make them useful for jewellery, wires, building and even artificial knee joints!

Alloys

Most of the metals we see every day are not pure metals. They are often mixtures. Alloys are mixtures based on metals. An alloy often has better properties than the pure metal. Steel is an alloy of iron and carbon that is stronger than pure iron. Brass is an alloy of copper and zinc that has a better shine and is stronger.

8 List the properties needed for each of these uses:

a the wire on a set of headphones

b the jaws of a bolt cutter

c a column to hold up a building

d a central heating radiator.

▲ Metals are used to make jewellery

▲ Copper is used in electrical wires because it conducts electricity well

▲ Metal girders are used in buildings

▲ Metals are used in branding irons

Aluminium is a useful metal because it is light and strong. Aluminium also does not corrode quickly.

Copper is soft and easily pulled into wires. It conducts electricity very well. Most large modern buildings have miles of copper wiring running through them. The water pipes feeding your taps are probably made of copper as well.

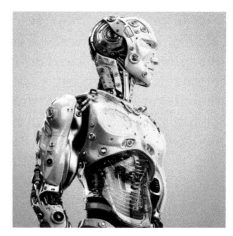

▲ Metals are very tough - just look at this robot!

▲ Aluminium is often used to cover large spaces in modern buildings

Why so tough?

What makes metals so tough? The atoms in metals are arranged in giant structures with all the atoms linked together. This makes the metal very strong. Everything is held tightly in place.

When metals are heated, the structure begins to break down and the metal gets softer. Sculptors and blacksmiths use this to make the metal easier to beat into shape.

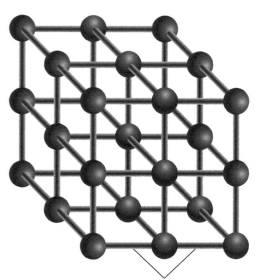

These strong bonds hold the metal atoms in a rigid shape.

▲ Metallic structure

✖ Carbon

Even a diamond so small that you could fit it in the middle of this letter 'o' costs about £500! The really big diamonds are impossible to price.

Diamonds are made of carbon. In a diamond, all the carbon atoms are tightly linked by chemical bonds.

The diamond is a giant crystal of atoms. Because they are linked so tightly and in such a rigid structure, diamond is very tough. A diamond can cut through glass easily.

Next time your dentist drills into a bad tooth, the tip of the drill might have diamond coating on it.

▲ Marilyn Monroe sang that diamonds are a girl's best friend. Well they are certainly good to have around!

▼ **Each atom is at the centre of four links**

▼ **The structure of diamond**

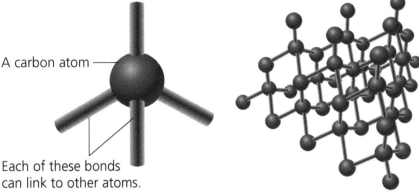

A carbon atom

Each of these bonds can link to other atoms.

▼ **The structure of graphite**

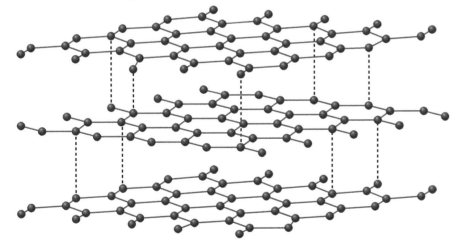

Graphite is also carbon. But graphite is soft and slippery. It is used as a lubricant to help surfaces slide over each other. In graphite, the carbon atoms link together strongly in sheets but there are only a few links between the sheets. This means that the sheets slide over each other.

Graphite is also used in some pencils.

Episode 6 Plastics

Yumi lives in Tokyo. She is a fan of things to eat with, if they are made of plastic. If you can use it to put it in your mouth and it's made of plastic Yumi collects it!

1 What would you like to collect? Why?

2 What is the strangest collection you have ever heard of?

Why plastic?

Plastic is an artificial substance that is made by joining together small molecules into long chains. The small molecules are called the monomers and the long ones are called polymers. Ethene is a simple chemical that can be joined in chains to make polythene. Styrene is another simple chemical that can be joined together to make polystyrene.

▲ Yumi has the world's biggest collection of plastic spoons, knives, forks, chopsticks, cocktail sticks and tongs

▼ **Monomer and polymer molecules**

Polymerisation

Monomers

Polymer

▲ Polythene can be used to make plastic bags or even this drain in Mexico

▲ This white packaging is polystyrene. Air was blown into it to make it bubble up when it was made. If you do not add those bubbles, polystyrene can be shiny and brittle

Plastic properties

All plastics are waterproof, do not corrode like metals, are lightweight and can be moulded in many shapes. This makes them very useful.

Plastic problems

Most natural things break down in the environment. Microorganisms eventually reduce even the largest trees to simple chemicals. All living things and the things made from them are biodegradable.

▲ **Plastics do not break down in the environment**

Most plastics are not biodegradable. That means that plastic in the environment can last forever. Damage done by the environment can break the large lumps of plastics down into tiny balls of plastic, but nothing can complete the job. These tiny lumps of plastic have been found all over the world.

3 Write a sentence containing the word 'biodegradable'.

4 Give one advantage of plastics not being biodegradable.

5 Give one problem caused by plastics not being biodegradable.

6 We are encouraged to recycle plastic waste nowadays. Give two advantages of recycling.

Summary

Properties of metals

» Metals are strong, can be polished to a shine, conduct heat and electricity well. They have high melting points and are solids at room temperature.

» Gold is so unreactive that it can be found as the pure metal in the ground. Most metals are found combined with other chemicals to form a mineral ore.

» Metals are used for electrical wires, water pipes, girders to reinforce buildings, to make cars and other vehicles and even space rockets.

The periodic table

» The periodic table lists all the elements that exist. It organises them into groups that have similar chemical properties.

» Most of the elements in the periodic table are metals. They are shiny, tough, can be hammered into flat sheets or pulled into thin wires. They conduct heat and electricity well.

» A small number of elements at the top right-hand side of the periodic table are non-metals. These are very different from each other and from metals. Some are gases, some are solids. They do not conduct heat or electricity well. They cannot be polished to a shine or made into sheets or wires.

» Group 1 metals are very soft and react very quickly with water or oxygen. They burn very easily.

» Group 7 are non-metals and include chlorine and iodine, which are used as disinfectants.

Useful materials

» Metals have many properties that make them useful. Aluminium is strong and light and does not corrode easily, so is used in buildings. Copper is easy to make into wires and conducts electricity and heat well. It is used to make electrical cables and pipes for carrying water.

» Carbon is an element that can exist in two forms: diamond or graphite. Diamond is very hard because each atom is held in place by four neighbours. Graphite is slippery because the atoms form flat sheets that slide over each other.

» Metals are found in compounds in the ground called minerals. The minerals are often mixed in with rock to make an ore.

» Chemical reactions are used to extract the metal from the mineral. Heating an ore with carbon is often used to extract the metal. Extracting metal is expensive. Mining can cause environmental problems. However, metals have many uses in the modern world.

» Plastics are artificial compounds made when simple molecules called monomers join together in long chains to make a polymer. Polythene and polystyrene are both examples of plastics.

» Plastics are light, waterproof, do not corrode like metals, and can be moulded into any shape. The are used for many different purposes, from plastic bags and drainpipes, to packaging foam and plastic cutlery.

» Plastics are not biodegradable and fragments of plastic can now be found all over the world.

Show you can...

A Have a discussion about the difference between elements and compounds.

B Draw a poster showing the key properties of three metals. Your poster should also show how these properties affect the way the metal is used.

C You work on a science helpline. Record definitions for each of the following words: element, atom, compound, metal and non-metal.

D Prepare a paper showing the advantages and disadvantages of opening a mine in your town. You must present both sides of the argument fairly.

E Make an exhibition for a science museum, which includes a display of as many types of plastic items as possible. Prepare a leaflet that shows why plastics are so useful and why they need to be recycled carefully and not just thrown away.

Episode 1 Acids and alkalis

Happy Holi!

At Holi, people throw coloured powders around to celebrate the triumph of good over evil. It is also a great excuse for a party. Luckily, the dyes wash out easily!

Some dyes stick and some even change colour. Indicators are dyes that change colour in acid. Many indicators are made from plants. An indicator from beetroot is dark red in acid solutions and yellow in alkaline solutions.

▶ **Holi is a Hindu festival that happens every March**

Off the scale

Chemists use the pH scale to describe how strong an acid or alkali is. Pure water has a pH of 7. Anything below pH7 is an acid. The lower the number the stronger the acid. Substances with a pH above 7 are alkalis. The strongest alkali has a pH of 14.

1 Write a sentence with the word 'indicator' in it.

2 Why are indicators useful?

Spot the acid (or alkali)

3 What is the pH of the solutions these strips have been dipped in?

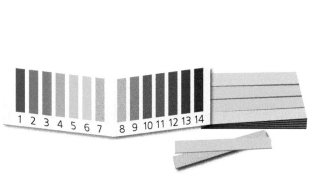

1 2 3 4 5 6 7 8 9 10 11 12 13 14

Lemon juice   Indigestion medicine   Limewater   Pure water   Bleach

▲ **A book of pH test papers. Just tear off a strip of paper and drop it in the test liquid. The colour change tells you the pH.**

Reactions of acids

Acids and metals

Some metals react with acid to produce hydrogen. Zinc is a good example. It produces bubbles of hydrogen as it dissolves. Zinc sulfate solution is left in the test tube.

zinc + sulfuric acid → hydrogen + zinc sulfate

magnesium + hydrochloric acid → hydrogen + magnesium chloride

Practical

To test for hydrogen
Light the gas with a burning splint.
Hydrogen burns with a popping sound and a blue flame.
It makes water vapour.

Safety note: if you're doing this practical you must wear eye protection.

▲ Hydrogen burns with a popping sound and blue flame

4 What would be formed when zinc reacts with hydrochloric acid?

5 What is made when calcium reacts with sulfuric acid?

6 You test a mystery gas by burning it. What two clues would you look for if you thought the gas may be hydrogen?

7 Gold is not affected by acid. Brass is made of copper and zinc, but it shines like yellow gold. How could you test these metal rings to see if they were gold or brass?

Practical

Making hydrogen
Do all the metals produce the same amount of hydrogen? Carry out an investigation to find out if zinc and magnesium produce the same amount of hydrogen when they dissolve in sulfuric acid.

Acids and carbonates

When acids react with chemicals called carbonates they produce a salt, carbon dioxide and water. The carbon dioxide bubbles off very quickly. This reaction is used in some fire extinguishers to shoot water out of the container as the gas expands.

Acids and alkalis

When acids react with alkalis they produce water and a salt. This reaction is called neutralisation. The word equation below shows what happens when hydrochloric acid and sodium hydroxide solution react.

hydrochloric acid + sodium hydroxide → sodium chloride + water

8 a What will happen when the plunger in the fire extinguisher below is pushed down?

b Explain how this will force water out of the extinguisher.

▼ **Fire extinguisher with calcium carbonate and dilute acid solution**

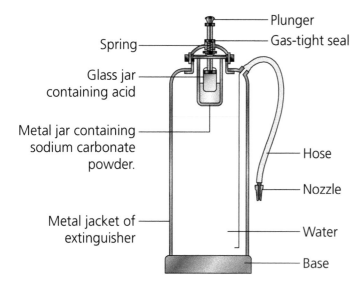

Plunger
Spring
Gas-tight seal
Glass jar containing acid
Metal jar containing sodium carbonate powder.
Hose
Nozzle
Metal jacket of extinguisher
Water
Base

Practical

Test for carbon dioxide
Bubble the gas through limewater. The limewater will turn milky white.

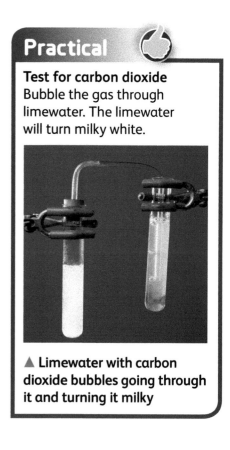

▲ **Limewater with carbon dioxide bubbles going through it and turning it milky**

Sodium chloride is the chemical name for ordinary table salt. To chemists, it is only one of many types of salts. Any reaction between an acid and an alkali produces a salt. We call the stuff we put on our fish and chips salt. We do not get our salt (sodium chloride) from a chemical reaction, but from the ground or the sea. Seawater is a solution of salt (sodium chloride) in water.

Making salts

When chemists want to make salts they often react an acid with a base. A base is a substance that makes an alkali when it mixes with water. Many metal oxides are bases.

Copper sulfate is a beautiful blue salt. Chemists make it by mixing sulfuric acid and copper oxide. The reaction is:

copper oxide + sulfuric acid → copper sulfate + water

▲ In hot countries, they let seawater into large flat ponds. The water evaporates in the warm sunshine to leave the salt behind. It is scooped up by a digger. Then more seawater is let in and the process starts again

▲ These crystals in a cave in Mexico are quite large!

Most of the crystals you will see will be much smaller than the ones shown in the photograph. What decides how big a crystal will be?

9 Estimate the size of the rock crystals in the photo. Look for the man at the bottom to give you a clue!

Practical

Crystal sizes
Who can make the biggest copper sulfate crystal? Investigate how the rate of evaporation affects the size of copper sulfate crystals produced.

Did the crystals form quickly or slowly? Give a reason for your answer.

Episode 2 Chemistry and energy

Warming up!

▲ You know what it's like. On a cold day you really need a warm coffee. Or maybe a roll around in the snow like these cold weather fans?

Hot drinks without flames

Combustion is a chemical reaction between a fuel and oxygen that gives out heat. We call it burning! But what if you want heat without a flame?

▼ When you press the button at the base of this can, it breaks a barrier and allows the chemicals to mix with the water and produce heat. Five minutes later, you have hot coffee!

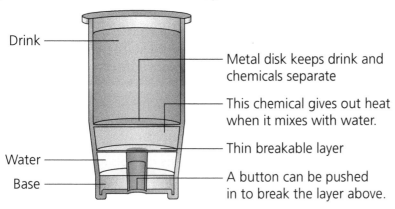

Drink

Metal disk keeps drink and chemicals separate

This chemical gives out heat when it mixes with water.

Thin breakable layer

Water

Base

A button can be pushed in to break the layer above.

Practical

A neutralising reaction
Neutralisation is another reaction that gives out heat. Investigate the change in temperature when you mix sodium hydroxide and hydrochloric acid solution.

Cooling down

If you twist your ankle it will start to swell. First aiders cover the damaged part with a cold pack, which cools the injury. This helps to stop the swelling.

▲ **A cold pack helps to stop the swelling on an injury**

▼ **How a cold pack works**

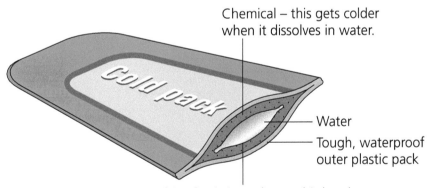

Chemical – this gets colder when it dissolves in water.

Water

Tough, waterproof outer plastic pack

Thin plastic inner bag – this breaks when you twist the pack.

In this cold pack, there are some chemicals that take in heat when they react with water. Normally these chemicals are kept apart by a barrier. When you twist the pack, the chemicals can mix and start to react. This takes in heat. Two minutes later, you have a cold pack for your twisted ankle!

Practical

Cool it!
When ammonium chloride dissolves in water it takes in heat. How cold can you get it to go? Investigate the drop in temperature with different amounts of ammonium chloride.

Episode 3 Speeding up

✳ Kill that acid!

▲ **You shouldn't have eaten that. Even for a dare! You have a stomach ache and it's bad**

Stomach ache is usually caused by too much acid in your stomach. Most indigestion medicines work by neutralising the acid.

Calcium carbonate reacts with hydrochloric acid. It produces calcium chloride, carbon dioxide and water. How can you speed up this reaction?

1 What is the scientific meaning of the word 'neutralise'?

2 How could a neutralisation reaction help with stomach ache?

Practical

Speedy stomach saver!
A group of researchers came up with three ideas:

- Hot acid will react more quickly than cold acid.

- Powder will react more quickly than lumps of calcium carbonate.

- Stirring the mixture will speed up the reaction.

Pick one and see if the researchers are correct. Does it speed up dissolving of the calcium carbonate?

Which of the researchers' ideas are not suitable for stomach ache medicines? Why?

How will the results for your investigation help you decide which stomach ache medicine to take?

✺ Hot and sticky!

When a surfboard is damaged, it is possible to patch any holes or tears with epoxy resin and glue. Just like repairing a bicycle tyre. You then have to wait while the glue sets to go hard.

The epoxy glue is soft before it sets. It is smoothed into place and then goes hard. To speed up the setting a chemical called a catalyst is mixed in with the glue.

A catalyst is a chemical that speeds up a chemical reaction. Catalysts are not changed by the reaction and are not used up.

▲ **Fancy a bit of surfing? It's a great way to keep fit and the shorts look cool! But surfboards are easily damaged, and not just by shark bites! Mike Coots had his leg bitten off by a shark when he was 17. But it has not stopped him surfing or campaigning to protect sharks**

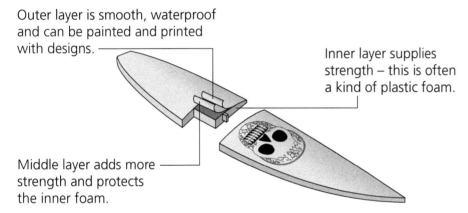

Outer layer is smooth, waterproof and can be painted and printed with designs.

Inner layer supplies strength – this is often a kind of plastic foam.

Middle layer adds more strength and protects the inner foam.

▲ **The surface of the board is made of layers of fabrics stuck together**

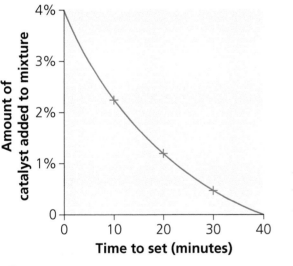

Amount of catalyst added to mixture

Time to set (minutes)

◀ **This graph shows how long it takes for the glue to go hard with different amounts of catalyst**

3 How does the catalyst affect the speed at which the glue sets?

4 a What are the advantages of quick-setting glue?

   b What might be one disadvantage of quick-setting glue?

5 Suggest one other way you could make the glue set more quickly. Give a reason for your suggestion.

Summary

✳ Acids and alkalis

» An acid is a chemical that turns litmus red. It has a pH of below 7.

» An alkali is a chemical that turns litmus paper blue. It has a pH above 7.

» Pure water is neutral and has a pH of 7.

✳ Reactions of acids

» Hydrochloric acid reacts with some metals to produce hydrogen and metal chlorides.

» Sulfuric acid reacts with some metals to produce hydrogen and metal sulfates.

» Hydrogen burns with a popping sound to produce water vapour.

» Acids react with alkalis to produce salts and water.

✳ Reactions and energy changes

» Some reactions give out energy and warm up their surroundings.

» Some reactions take in energy and cool their surroundings.

Rates of reaction

» Increasing the surface area of reacting chemicals, increasing the temperature of the reactants or their concentration makes reactions go more quickly.

» Catalysts increase the speed of a reaction.

» The speed of a reaction is measured by how quickly the reactants are used up or the products are made.

Show you can...

A Explain to a partner how you could test a mystery white powder to see if it was a carbonate.

B Create a poster to show how you can test for each of these gases: hydrogen and carbon dioxide.

C Prepare a presentation about the chemical properties of acids. Make your presentation interesting with lots of photographs or drawings.

D Write an account of how you could find out how temperature affects the rate of a simple chemical reaction, perhaps dissolving calcium carbonate lumps in dilute hydrochloric acid.

E Build and test a machine that could test the strength of a glue. How could a researcher use your machine to find out how quickly a glue sets?

Episode 4 Fuel and fires

Up in flames

▲ Fire triangle

To make a fire you need a fuel. This is something that can burn to give out light and heat. The light and heat are useful to us.

Combustion

Chemists call the chemical reaction when a fuel burns combustion. In a combustion reaction, chemicals in the fuel react with oxygen in the air.

The chemicals in the fuel are oxidised. This means oxygen is added to them. The fumes that come from a fuel are called oxides.

Coal is a good example of a fuel. It is mainly carbon.

carbon + oxygen → carbon dioxide

Natural gas is methane. Methane is a compound of two elements.

methane + oxygen → carbon dioxide + water

Practical

The best burn?
There are many possible fuels. But which one is best? Investigate a variety of different fuels to see which is the best one. Explain how you chose the winner in your fuel tests.

▲ Is this the most dangerous job in the world? Of course not! The stunt performers know exactly how fire burns. They can control it and make these action shots safe. But don't try it at home!

1 a What three things do you need to start a fire?

b What is given out when something burns?

2 List the things you use a flame for at home.

3 What fuels do you burn at home?

4 What is a combustion reaction?

5 What two elements are present in methane?

Smoking is bad for you!

Some fuels burn very cleanly and so do not produce much smoke. Others produce a lot of dark smoke. The amount of smoke depends on the fuel and the amount of oxygen available.

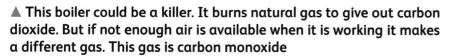

▲ This boiler could be a killer. It burns natural gas to give out carbon dioxide. But if not enough air is available when it is working it makes a different gas. This gas is carbon monoxide

▲ Fires give out heat and light. They also give out fumes or smoke

Carbon monoxide is poisonous. People have been killed because the boiler they used didn't work properly and produced carbon monoxide. Boilers, like all gas appliances, need to be checked regularly to make sure they are safe.

Smells bad!

In 2012, four million people died through inhaling the smoke from cooking stoves and fires. The worst stoves are very inefficient and produce lots of smoke. These stoves can fill a home with smoke very quickly, especially if they do not have a chimney to let out the fumes.

6 a What might make a boiler produce carbon monoxide instead of carbon dioxide?

  b Why is carbon monoxide dangerous?

▲ The smell of grilled bacon is great, but the smoke that comes off a wood fire is not so good

Nitrogen oxides:
these gases dissolve in rainwater to make acids.
They irritate the lungs and can cause asthma attacks.

Carbon dioxide:
a colourless gas that dissolves in
rainwater to make a weak acid.

Sulfur dioxide:
a gas that makes a strong acid
when it dissolves in water. Very
damaging to the lungs.

Carbon particles:
tiny specks of carbon that can cause
asthma, cancer and other lung diseases. The
largest particles settle out of the air as soot.

Carcinogens:
these chemicals
cause cancer.

▲ What's in smoke?

Smart cooker

The cards below give you information about some stoves that are used around the world.

Activity

Chirpy chimney

Investigate the best sort of chimney to take away smoke from a fire. Is tall and thin better than short and fat?

Activity

The best cooker
Pick one of the situations below. Design the best possible cooking stove for this situation. Present your design as a diagram with labels that explain all its key features.

A flat in central London

A farmhouse in North Wales

A campsite in Cornwall

Improved stove

Fuel	Wood, straw or grass
Fumes	Very few, removed by chimney
Country	Nepal
Environmental impact	Medium, fuel is more sustainable since stove makes efficient use of fuel

Gas cooker

Fuel	Natural gas
Fumes	Very few fumes when the cooker is working properly
Country	England
Environmental impact	Extracting gas needs a lot of energy. Gas is a fossil fuel so increases global warming

Clay stove

Fuel	Wood
Fumes	Lots of fumes
Country	Bangladesh
Environmental impact	Medium, fuel is more sustainable since stove makes efficient use of wood

Camp fire

Fuel	Wood or straw
Fumes	Lots, especially if the wood is wet
Country	Worldwide
Environmental impact	High, the wood can be gathered locally but a camp fire uses a lot of wood to cook food

Episode 4 Fuel and fires

Big oil

In 2014, Saudi Arabia sold crude oil worth over £255 billion! That's enough to buy every adult in Wales a brand new Ferrari (but they'd have to buy their own petrol!).

Oil is a fossil fuel. It has been created over millions of years by layers of rock pressing down on dead plants and animals.

The buried plants and animal remains are slowly converted to thick, black crude oil. When people drill down through the rock on top the oil can get to the surface.

7 What is a fossil fuel?

8 Why are fossil fuels called non-renewable fuels'?

▲ Sometimes the pressure on the oil is so great that it squirts out of the drill hole all by itself. This is an oil gusher

Crude oil is a mixture of different hydrocarbons. A hydrocarbon is a chemical containing only hydrogen and carbon. Fractional distillation separates out these hydrocarbons according to their boiling points.

Crude oil also contains small amounts of sulfur. This can cause problems when the oil burns because it forms the acid gas sulfur dioxide. Oil refineries try to remove the sulfur from the crude oil. The sulfur can be sold to chemical companies to make sulfuric acid.

▼ **Fractionating column with useful products labelled**

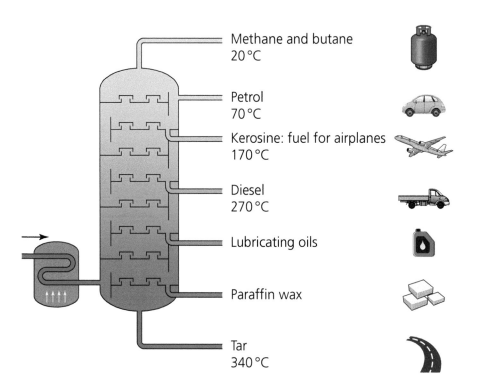

Methane and butane
20 °C

Petrol
70 °C

Kerosine: fuel for airplanes
170 °C

Diesel
270 °C

Lubricating oils

Paraffin wax

Tar
340 °C

Many of the chemicals produced by an oil refinery can be used to make other things. Most plastics come from oil.

9 What does a fractionating column do to crude oil?

10 a Which part of crude oil has the lowest boiling point?

b Which part of the fractionating column does this part come out from?

11 What is the problem with sulfur in the crude oil?

12 Suggest three ways people could reduce the amount of oil they consume.

✳ Big stink

Beijing is the capital city of China. There are millions of cars on its streets. They use petrol and diesel. Factories and homes also need to burn fuel for energy. All this produces a lot of smoke.

Damp weather produces fog. Fog is tiny droplets of moisture in the air. When smoke and fog mix they make smog.

13 What is in smog?

14 a Which month is the worst for pollution in Beijing?

b Suggest a reason why this month might be so bad.

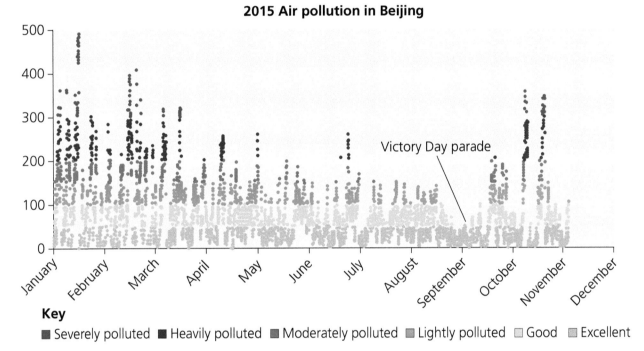

2015 Air pollution in Beijing

Victory Day parade

Key

■ Severely polluted ■ Heavily polluted ■ Moderately polluted ■ Lightly polluted □ Good ▢ Excellent

▲ **Smog in Beijing**

In August 2015, the Chinese government closed factories around Beijing and banned many cars from the streets. After a few days, the smog cleared. The people could see blue skies for a parade to celebrate the end of the Second World War. But the factories were started up afterwards and the pollution was as bad again by September.

▲ **The pollution got worse once the factories started up again**

Episode 5 Air and water

✷ Fit to breathe?

Mars One

This is a project to put a human settlement on Mars by 2030! The trip to Mars will be one-way. There will be no coming back for the volunteers. Would you be willing to leave Earth to go into space and never come back?

The air around a planet is called its atmosphere. Gravity holds this atmosphere close to the planet's surface. Mars is a smaller planet than Earth so there is less gravity. Most of the Martian atmosphere just escapes into space.

▲ **The Mars One settlement**

Table 1 Comparing Earth and Mars

	Earth	Mars
Atmosphere	Nitrogen (77 %) Carbon dioxide (0.04 %) Oxygen (21 %)	Nitrogen (3 %) Carbon dioxide (96 %)
Distance to Sun	149 597 891 km	227 936 637 km
Gravity	2.66 times that of Mars	0.375 times that of Earth
Tallest volcano	Mauna Loa in Hawaii 6.3 miles high	Olympus Mons 26 km high
Hours in a day	Just slightly under 24 hours	24 hours, 37 minutes
Days in a year	365 days	687 Earth days
Polar caps	Permanently covered with frozen water (ice)	Covered with a mixture of frozen carbon dioxide and water
Surface temperature	14° C	-63° C
Moons	1	2

1 Give two differences between Mars and Earth that will make it difficult to survive on Mars.

2 How far is Mars from the Sun?

3 How many days are there in a Mars year?

Earth's atmosphere

Mars is a bit like Earth was billions of years ago. It has a very thin atmosphere and very little water. It is not a good place to live! Volcanoes and algae made the Earth like it is today.

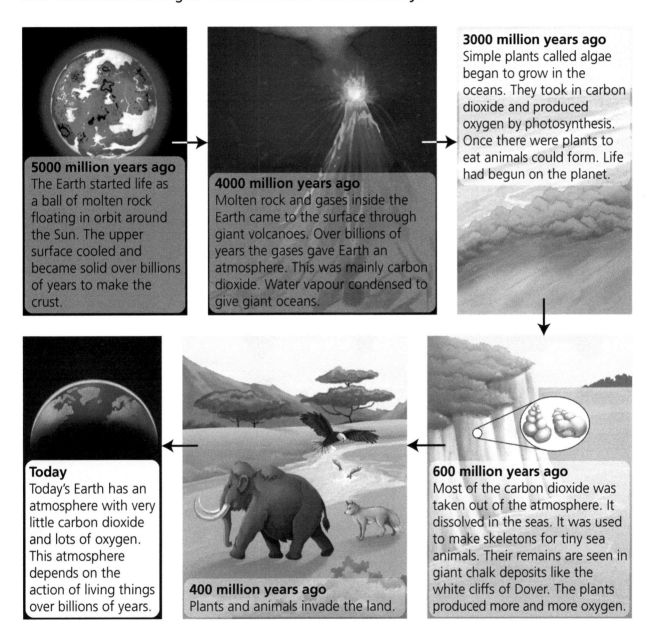

5000 million years ago
The Earth started life as a ball of molten rock floating in orbit around the Sun. The upper surface cooled and became solid over billions of years to make the crust.

4000 million years ago
Molten rock and gases inside the Earth came to the surface through giant volcanoes. Over billions of years the gases gave Earth an atmosphere. This was mainly carbon dioxide. Water vapour condensed to give giant oceans.

3000 million years ago
Simple plants called algae began to grow in the oceans. They took in carbon dioxide and produced oxygen by photosynthesis. Once there were plants to eat animals could form. Life had begun on the planet.

Today
Today's Earth has an atmosphere with very little carbon dioxide and lots of oxygen. This atmosphere depends on the action of living things over billions of years.

400 million years ago
Plants and animals invade the land.

600 million years ago
Most of the carbon dioxide was taken out of the atmosphere. It dissolved in the seas. It was used to make skeletons for tiny sea animals. Their remains are seen in giant chalk deposits like the white cliffs of Dover. The plants produced more and more oxygen.

Practical

How much oxygen?
Elodea is a simple water plant. It takes in carbon dioxide and gives out oxygen by photosynthesis. How long does it take to make 1 ml of oxygen? Carry out an investigation to find out.

4 When did plants first appear on the Earth?

5 Why does the Earth's atmosphere contain oxygen?

The greenhouse effect

The Eden Project in Cornwall has several giant greenhouses shaped like domes.

Light goes into the dome through the clear plastic roof. The dome walls keep the heat in. They can grow bananas and sugar cane in Cornwall!

The atmosphere of the Earth acts like the walls of the domes at the Eden Project. A layer of gases high above the planet reflects heat back to the planet's surface. This keeps us warm.

▼ **How a 'greenhouse' Earth works**

No greenhouse – a lot of the energy that reaches the Earth from the Sun is reflected back in to space.

On a 'greenhouse Earth' the gases in the atmosphere reflect the energy back to the surface.

Over the last 100 years, the Earth has been getting warmer. This is called global warming. Most scientists think this warming is due to increases in the greenhouse gases. The main greenhouse gas is carbon dioxide. This is produced every time fossil fuels are burnt.

Episode 5 Air and water

102

▼ **Graph to show rise in global temperature**

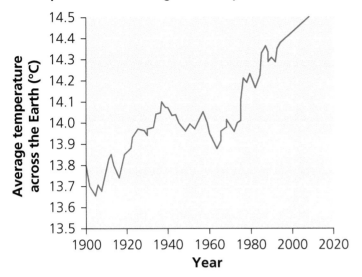

▼ **Graph to show rise in CO₂ levels**

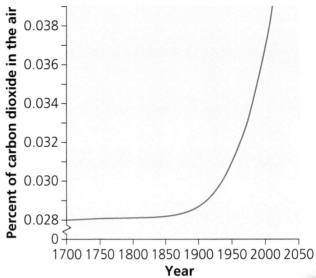

Many scientists believe that Global warming leads to climate change. The warming of the atmosphere can also change weather patterns. Britain will probably have warmer, wetter winters but summers that are dryer and hotter.

Storms will also become more common. The number of storms has nearly doubled between 1985 and 2015. Floods devastated Somerset in 2014 and the north of England in the winter of 2015. They were both thought to be very rare events, but are now expected more often due to climate change.

6 What does the phrase 'global warming' mean?

7 Give one piece of evidence that suggests carbon dioxide levels in the atmosphere and global warming are linked.

8 What problems might be caused by warmer, wetter winters and hotter, dryer summers in the UK?

▲ Floods hit parts of England in 2015

▲ Burning fossil fuels in vehicles produces carbon dioxide

▲ Burning coal or oil in power stations to make electricity produces carbon dioxide

▲ Methane is produced from decaying rubbish at landfill sites

▲ Cattle produce methane in their guts

9 a What is the connection between fossil fuels and carbon dioxide in the atmosphere?

b What greenhouse gas is made when rubbish decays?

c Why might cows be causing global warming?

d How has human activity helped to increase greenhouse gases in the atmosphere?

✖ Fancy a drink?

▲ **What has a reservoir in mid-Wales got to do with a bomb that can bounce on the water?**

In the 1900s Birmingham was growing very quickly. But it did not have enough clean water to drink. So engineers built dams in the Elan valley in mid-Wales to make a reservoir. These store the water running off the hills. It can then be passed to Birmingham along a giant aqueduct.

During the Second World War, Barnes Wallace worked on a bomb that bounced on the surface of the water. This bomb was designed to destroy huge dams in Germany. The bombs were tested in the Elan valley on a small dam called Nant-y-gro, which was built to provide water for the workers as they built the big dams in the valley.

10   What is a reservoir?

11   Why is it not a good idea to drink water straight from a reservoir?

▲ **A Lancaster bomber – these planes carried the bouncing bombs**

Water processing

Water from a reservoir may have mud, bits of plants, small animals and microorganisms in it. Not good to drink!

▼ **Making water safe to drink**

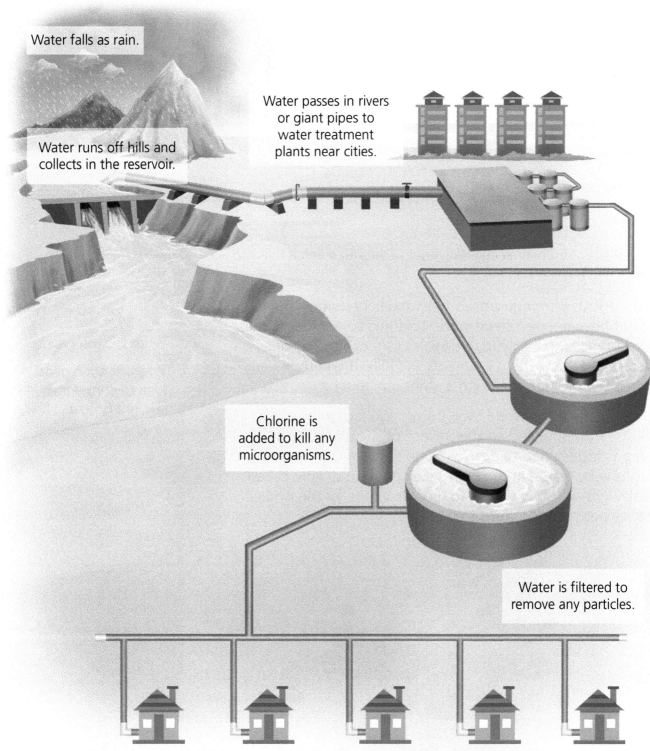

Water falls as rain.

Water runs off hills and collects in the reservoir.

Water passes in rivers or giant pipes to water treatment plants near cities.

Chlorine is added to kill any microorganisms.

Water is filtered to remove any particles.

Pipes carry the water to homes.

Practical

Clean water?
Millions of people around the world cannot just turn on the tap to get clean water. They depend on water from rivers or lakes, which can be badly polluted. Build and test a simple water filter. Try to convert muddy water into clear water.

What else would you need to do before the water was safe to drink?

Bottled or tap?

Bottled water can be a thousand times more expensive than water from the tap. Companies that sell bottled water often say it contains 'more minerals' that give it a better taste.

Tap water is safe to drink. It has been treated to remove the bits of dirt and most of the microbes. This is done in water treatment plants. It still contains small amounts of dissolved substances but these are safe to drink.

▲ **Can you tell the difference between bottled water and tap water?**

Practical

Testing the taste
Carry out a test to compare tap water with bottled water. Can you detect any differences between them?

12 How could you test to see if bottled water did contain more dissolved minerals than tap water?

Summary

Combustion

» Burning fuels give off carbon dioxide, water (vapour), and oxides of nitrogen. Some fuels also contain sulfur and these give off sulfur dioxide as well when they burn.

» When fuels burn without enough air, they produce carbon monoxide. Carbon monoxide is a poisonous gas.

» Soot is made up of tiny particles of carbon. It can be made when fuels burn in a limited supply of air. Some diesel engines produce particles that are so small that they are invisible. These can cause problems when they get into our lungs. Slightly larger particles in the upper atmosphere can make the light from the Sun dimmer.

» Nitrogen oxides and sulfur dioxide are acid gases. When they dissolve in rain, they make acid rain. This can damage buildings and cause health problems for humans.

» Crude oil is a mixture of a very large number of compounds. Crude oil is found in deposits underground, for example the oil fields under the North Sea.

» Fractional distillation separates crude oil into fractions depending on their boiling point. Different fractions have different uses.

» Oil refineries separate crude oil into fractions and then treat each of these to make useful products such as petrol and diesel fuel, lubricating oils and gases.

The atmosphere

» The Earth's first atmosphere was made by gases given out by volcanoes. Water vapour in these gases condensed to give large oceans. This early atmosphere was mainly carbon dioxide with little or no oxygen.

» About 3 billion years ago, algae and plants developed. These produced oxygen, and took in carbon dioxide by photosynthesis. Photosynthesis can be represented by the word equation:
carbon dioxide + water → glucose + oxygen

» Carbon dioxide was removed from the early atmosphere by dissolving in the oceans and by photosynthesis. Most of the carbon from this carbon dioxide is now locked up in rocks as carbonates and fossil fuels. The Earth's atmosphere is now about four fifths (80%) nitrogen and about one fifth (20%) oxygen, with small amounts of other gases, including carbon dioxide, water vapour and argon, which is a noble gas.

» Carbon dioxide and methane are examples of greenhouse gases. Increased levels of greenhouse gases in the atmosphere cause the temperature to increase.

» Burning fossil fuels returns carbon dioxide to the atmosphere. Decay of rubbish at landfill sites produces methane. Cattle also produce methane from their gut.

Water to drink

» Water that is safe to drink has small amounts of dissolved substances and low levels of microbes. Most drinking water is produced by choosing a suitable source of fresh water, filtering to remove solids and sterilising to kill microbes.

» If supplies of fresh water are limited, salty water can be distilled to produce fresh water. This requires a large amount of energy.

Show you can...

A Explain to a partner the features you need in a good fuel. Give reasons for your choices. Agree with your partner which fuel you think is best to use to: provide energy for a moving vehicle, cook a meal, heat a home.

B Create a poster to show the way the Earth's atmosphere has changed since the planet first formed.

C Prepare a 2-minute radio script to explain the difference between global warming and the greenhouse effect. The item should be suitable for ordinary members of the public who may not understand the science behind these terms.

D Write a paper to show how your home or school could reduce their use of fossil fuels. Suggest at least three actions the school could take and give reasons for each one.

E Build and test a fuel brick. This should be clean, easy to light, give out a lot of heat and last a long time.

Physics 1

Episode 1 Full of energy

✳ Saving Iron Man

What ways are there to store energy?

▲ Iron Man must have a good system to store energy. Otherwise he would have to stay connected to the wall socket all the time

▲ How do you think light sabers might store energy?

▲ The Sun is a huge store of energy for us. But how easy is it to collect?

▲ This cyclist stores energy as he goes downhill. He knows he'll find it easier to go up the next hill that way!

▲ This old clock stores energy in a tightly-wound spring to move the hands round the face

▲ The pulled string on this bow stores energy to shoot the arrow forward when it is released

▲ The twisted rubber band in this toy aeroplane makes the propeller go round and keeps the model in the air

Practical

Catapults

A catapult is a simple toy to throw a marble over a short distance. The energy stored in the stretched elastic transfers to the marble and pushes it forward. Plan and carry out an investigation to see how the distance you pull back the elastic affects how far your marble goes.

1 List at least five different ways to store energy.

2 What type of energy store do the following depend on?

   a A Formula 1 racing car.

   b A new smartphone.

   c A marathon runner.

▲ This stone has stored heat from the oven. When they put it on your table you can cook your steak just how you like it!

Energy: what's it like?

Energy is strange stuff. We cannot see it, hear it or smell it. We can only detect it when it makes something happen.

Energy is a bit like money. It is great to have lots of money in the bank. But money is only useful when you spend it! So, energy is only noticeable when some leaves the store.

Energy cannot be created or destroyed – it is just transferred from one form to another.

We can show these changes as an energy transfer diagram.

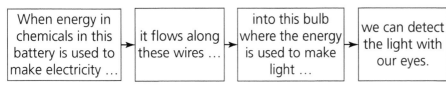

| When energy in chemicals in this battery is used to make electricity … | → | it flows along these wires … | → | into this bulb where the energy is used to make light … | → | we can detect the light with our eyes. |

▲ **Energy transfer from a battery to a light bulb**

Energy forever

Energy stored in the gunpowder in fireworks is transferred into sound, light and heat when they go off. But what is left afterwards? Is all the energy used up?

In fact, the total amount of energy after the explosion is the same as before. Before, most of the energy was stored in the gunpowder. Afterwards it is in the light and heat and sound.

Light, heat and sound spread out. The energy is spread so thinly it cannot be detected. We say the energy has dissipated. However, the total amount of energy will still be the same.

4 Write a sentence with the word 'dissipated' in it.

5 Why might some people think energy is used up when it is transferred?

3 Draw an energy transfer diagram to show how energy transfers when:

a a kettle boils water for tea

b a solar cell makes electricity

c a drummer hits the skin of the drum.

▲ **Fireworks are fun energy transfers**

✳ Efficiency

Some energy transfers work very well. Most of the energy makes something useful happen. These energy transfers are efficient. But sometimes the transfer is inefficient. In inefficient transfers too much of the energy does something that is not very useful. We say that the energy has been wasted.

Table 1 Efficiencies of light sources

Light source	Efficiency %
Candle	0.04
Traditional lightbulb	17
Halogen bulb	24
LED	10
Fluorescent tube	13

6 What is the useful energy output of a light bulb?

7 What is the wasted energy for a light bulb?

8 Which of the lights shown in Table 1 is the most efficient?

▲ The efficiency of each of these lights will be different

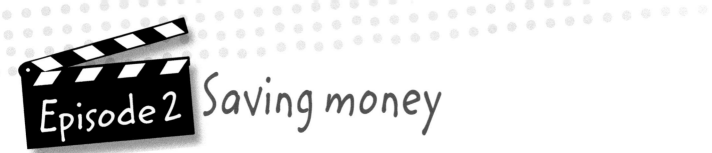

Episode 2 Saving money

✳ Where does the energy go?

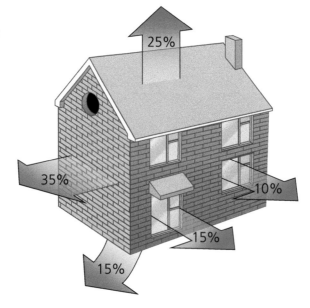

Hi! My job is to make sure that your heating bills are low. I do that with lots of lagging. We call it insulation.

The energy to heat your home may come from electricity, gas, coal, wood, or even the sunlight coming through big windows. The problem is that your home leaks! You get it nice and warm and then heat leaves through windows, walls, open doors and even into the cold ground.

Heat always flows from where there is more heat to where there is less heat.

▶ **Diagram of heat loss from home**

25%

35%

10%

15%

15%

1 List the ways you heat your home.

2 Where is most heat lost from your home?

Cooling coffee

You can buy coffee in every station and on every street nowadays. The coffee is served in a paper cup so that you can drink it as you walk to work. But on a cold day your coffee will cool quickly. And cold coffee is no good.

▲ Coffee will cool down quickly when it is surrounded by cold air

Moving energy

Heat energy always moves from where there is more to where there is less. You know this because a cup of hot coffee gets cold more quickly in the winter. Think about ice cream. In the summer, the heat in the air and the sunlight makes it melt and dribble down your fingers if you do not eat it quickly enough!

Practical

The best cup?
What affects how quickly coffee cools in a paper cup? Plan and carry out a test to see what affects the rate of cooling.

◀ This is Chen's house in Kuching, Malaysia. Here it is much hotter outside than they want it to be inside. Air conditioning keeps the inside cool

3 Will heat leak into Chen's house or out of it?

4 Would insulation help to reduce Chen's bills for air-conditioning? Why?

Energy leaks at home

Insulation reduces the speed heat passes through something. It keeps heat in your house. If you do not have 30 cm of lagging in your loft you are wasting money!

Table 1 Energy saving in the home

Heat saving tip	Cost to install (£)	Saving per year (£)
Insulate your loft	400	140
Draught-proof your doors and windows	120	35
Double glaze your windows	3000	100
Add cavity wall insulation	450	160
Add underfloor insulation	1500	55

5 Which type of insulation saves you the most money in a year?

6 If you had £500 to spend to insulate your home, what would you do? Give reasons for your choices.

Best walls

Different substances let heat pass through at different speeds. The speed at which heat passes through the substance is called its thermal conductivity. Substances with low thermal conductivity do not let heat pass easily. We call them insulators.

▲ This mud wall in Morocco is nearly half a metre thick

▲ This castle wall in Scotland is made of stone and is 18 cm thick

▲ These refugees can only afford a thin sheet, of metal for the walls of their home

▲ This house in Cornwall has walls made of hay bales, that are then covered with a kind of concrete. It is over a metre thick!

▲ A yurt is made of thick woollen fabric stretched over a frame

7 Write a sentence with the words 'thermal conductivity' in it.

8 Which of the houses shown on these pages will heat up and cool down most quickly? Give a reason for your choice.

9 Which two houses will lose heat to the outside?

Activity

Do thicker walls mean thinner bills?

Walls let heat move through them. Plan an investigation to see how the thickness of the walls affects how easily heat can move through.

✳ Power stations

How much energy do we use?

Everything we do needs energy. Where does it all come from?

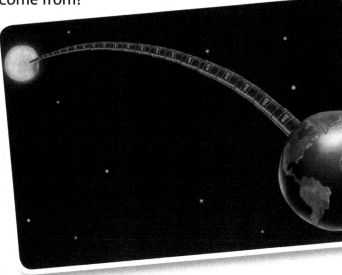

We use 1041 million tonnes of coal every year. If you imagine that as a single lump of black coal it would fill Trafalgar Square in London and be twice as tall as Nelson's column!

We use 43 billion barrels of oil every year. If one day's worth of barrels were stacked on top of each other they would reach the Moon and back (130 times!).

We use 1245 billion cubic metres of gas every year. If that was in one balloon it would measure 33 km across. Its shadow would cover all of London!

Non-renewable energy sources

Most of our energy comes from fossil fuels. These are fuels made from dead plants and animals buried millions of years ago. One day we will run out of these fuels. Fossil fuels are called non-renewable energy sources.

▲ No new fossil fuels are being made. We have to look in more and more difficult places to drill for oil and gas

Nuclear power uses heat from radioactive metals to boil water to make steam. The steam spins turbines.

▲ Heat from radioactive decay boils the water to make the steam to drive these huge turbines and create electricity

Nuclear power depends on the radioactive metal uranium. Uranium will run out as it is used up in reactors.

1 Why are coal, oil and natural gas called fossil fuels?

2 What does the word non-renewable mean?

3 List three non-renewable energy sources.

Renewable energy sources

We can also use other energy sources to make electricity. None of these energy sources will run out. These are renewable energy sources.

▲ Solar cells use energy in sunlight to make electricity. By 2020, almost 4 million homes in the UK will get their electricity from solar cells. Solar panels collect heat in sunlight to warm water. So you can get showers and baths without having to pay for the fuel to warm the water!

▲ Wind turbines use the energy in moving wind to turn a generator. The generator produces electricity. 10% of the UK's electricity comes from wind power

▲ Turbines in dams turn when water flows through them. The turbines turn generators to make electricity. Cruachan power station in Scotland has produced clean electricity for Glasgow every year since 1965

▲ Radioactive decay heats water deep underground to make steam. The steam can then be used to drive turbines to make electricity for Lhasa in Tibet.

4 List three renewable energy sources.

5 What do you use in school that runs on solar power?

6 List the advantages of renewable energy sources.

7 Give two disadvantages of wind farms.

Activity

The best windmill
Does a windmill with three blades spin faster than one with two? Plan an investigation to find out.

▲ The first biofuel was the wood that cave dwellers burnt. Nowadays biofuels include anything that we grow to make fuel. Biofuels can be solids or liquids. Some biofuels are used to make electricity in power stations

Chip shop van

Some chip shop vans run on diesel. But this diesel includes some old vegetable oil. Vegetable oil is used to fry food. Old vegetable oil is collected and treated to make into diesel fuel. The fuel can then be used to drive the van around town to sell more chips!

8 a Where does normal diesel come from?

b Is normal diesel fuel renewable or not? Why?

c What is special about the bio-diesel in some chip shop vans?

d Is the vegetable oil used to make diesel renewable or not? Why?

Summary

✳ Energy and efficiency

» We can only detect energy when it transfers from one form to another. Then we can detect the transfer of energy between energy stores.

» Energy is transferred when an object moves, or speeds up or slows down, or hits a barrier, or when something gets hotter.

» The total amount of energy before and after any transfer is always the same but the energy might be present in different stores.

» Energy transfers are not perfectly efficient. Some energy always spreads out into the system and is 'wasted', usually as heat.

» Insulation in houses reduces loss of energy by slowing down the speed at which heat can pass through different areas.

» The loss of heat through a wall depends on the thermal conductivity of the material that the wall is made from, how thick the wall is and the temperature difference between the inside and outside.

✳ Energy resources

» Energy resources are stores of energy that we use to do something, for example heat and light our homes, cook food, transport people and things around, or make electricity.

» Most of our energy comes from fossil fuels. These include coal, oil and natural gas. They are called non-renewable because they will eventually run out.

» Biofuels are fuels made by living things. Wood is the most common biofuel in the world. Some types of petrol include alcohol made from fermenting sugar cane. Some diesel engines can run on old vegetable oil.

» Renewable resources include solar, wind and wave power. They can be used to generate electricity and will never run out.

» Solar power uses energy from the Sun. The light makes electricity when it hits special solar cells.

» Rising and falling tides can be used to drive turbines. Wave power uses movements in waves. Hydroelectric power uses water flowing downhill to drive turbines. Dams collect the water and store it until it is needed to drive turbines to make the electricity.

» Wind power uses the movement of air to turn turbines to generate electricity.

» Geothermal power uses heat from rocks in the Earth to heat water to make steam to drive turbines.

Show you can...

A Work with a partner to list all the places where you might see energy being transferred. Take turns to make suggestions.

B Produce a poster of the ways to save energy at home. Show which three will save the most energy.

C Record a radio show with three people phoning in to ask how they could save energy at home. Give at least two suggestions to each caller.

D Describe how you could test the efficiency of a solar panel for a house roof.

E Build and test a paper cup to keep takeaway coffee hot for as long as possible. Design a label for it to show the users why the cup is so good at keeping their drink hot.

Episode 4 The force

▲ This is a competition for pinball players. The winner walks away with $10 000!

▲ When the firing pin hits the ball it pushes it onto the table. Physicists call that push a force

Pinball machines are all about forces. You have to hit the ball bearing with just enough force. Too much force and the ball overshoots. Too little force and the ball never gets to the top.

Activity

Marble madness
Create a marble madness game made of a course for the marble to roll along. Find out what has the biggest effect on the speed of the marble – the slope of the launcher or the height you release it from. Use what you find out to become marble madness champion!

1 Write a sentence with the words 'force' and 'push' in it.

2 Suggest two other games or sports that depend on forces. Explain what forces are needed in order to play.

Special forces

Physicists recognise a number of forces. When one object touches another to push or pull it, the force is a contact force.

Sometimes the objects do not even touch. A magnet can push another magnet away even when they do not touch. This is an example of a non-contact force.

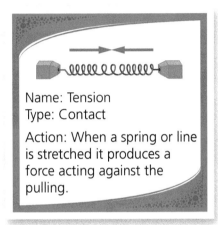

Name: Tension
Type: Contact

Action: When a spring or line is stretched it produces a force acting against the pulling.

Name: Air resistance
Type: Contact

Action: When something moves through air the air pushes against it.

Name: Friction
Type: Contact

Action: When two surfaces slide over each other a force tends to slow down their movement.

Name: Push
Type: Contact

Action: When a solid pushes directly on another solid.

Name: Gravitational
Type: Non-contact

Action: Gravity pulls all masses together.

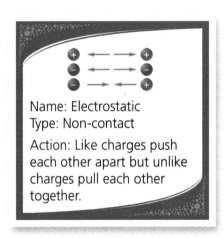

Name: Electrostatic
Type: Non-contact

Action: Like charges push each other apart but unlike charges pull each other together.

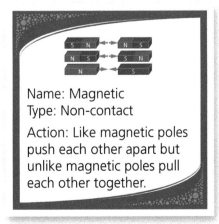

Name: Magnetic
Type: Non-contact

Action: Like magnetic poles push each other apart but unlike magnetic poles pull each other together.

3 What is the difference between the red and blue cards?

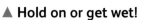

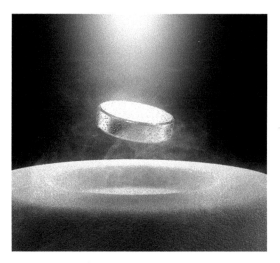

▲ Hold on or get wet! ▲ How is it floating?

4 Describe the forces you can see in these photographs. Are they contact or non-contact forces?

▲ It's bumper car mayhem!

▲ What stops the satellite from falling?

▲ Crazy or awesome?

▲ Would you do this?

Forces and work

Physicists use the word 'work' in a special way. Work is done when a force makes something move. So when you push open a door you do work on that door.

But if there is no movement no work has been done. So, if you push against a wall you do no work on the wall, even if you feel tired at the end!

In fact, the amount of work done depends on the size of the force and the distance you move the object. Stronger forces moving objects over longer distances mean more work is done.

▲ A baseball bat hitting a ball out of the stadium does work on the ball

▲ You do work when you wheel a wheelbarrow

▲ You do work when you lift weights

5 List three examples where a force does work on an object.

Friction and work

The back wheels of this dragster work very hard. They push against the ground and move the car along. The friction between the ground and the wheels produces heat. You can see this heat because the rubber tyres get so hot they start to give off smoke!

6 A drill bit gets hot when you drill into wood or metal. Where does this heat come from?

▲ Why does the drill bit get hot?

7 Suggest another place where friction can make something hot.

Episode 5 Road safety

✳ Speed kills

▲ Speed cameras are designed to deter people from breaking the speed limits on roads

▲ Sometimes, excessive speed is a factor in road traffic accidents

People are injured on Britain's roads every year. In 2013, 1713 people died. These numbers are one reason why we have speed limits on our roads. But do speed limits really help?

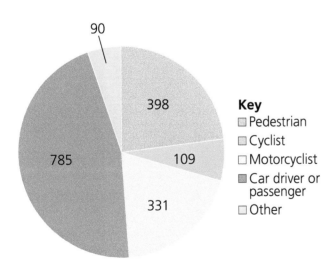

90
398
785
109
331

Key
☐ Pedestrian
☐ Cyclist
☐ Motorcyclist
☐ Car driver or passenger
☐ Other

▲ Deaths on UK roads in 2013

?

1 Do you think speed limits make our roads safer? Why?

2 How many cyclists were killed in 2013?

3 Suggest three ideas that you think would make Britain's roads safer for pedestrians and drivers.

Describing speed

You need two measurements to calculate speed. One is the distance travelled and the other is how long it took. Then you can put the data into this formula:

$$\text{speed} = \frac{\text{distance travelled}}{\text{time taken}}$$

The units for speed depend on what you use to measure the distance and time. Usually, in science, the units will be in metres per second. Speed limits in the UK are in miles per hour. In other countries they use kilometres per hour.

13 m/s = 30 mph = 48 kph

▲ This Formula 1 car goes at speeds that are much faster than those that would be safe on roads

4 A Porsche 911 goes from 0 to 60 in 4.6 seconds. What units is the 60 measured in?

5 If you measured how far a snail crawled in centimetres, and the time taken in minutes, what would your speed units be?

6 If a model car travels 12 metres down a slope in 4 seconds, what is the speed in m/s?

Safe stopping

If you see someone in front of you when you are driving a car, how long does it take to stop? There are two parts to the stopping distance:

* Thinking distance: how far the car goes before you react and press the brake.

* Braking distance: how far the car goes after the brake has been pressed.

The time you take to think before you press the brake is often called the reaction time. Most people react in a second or less. But cars can travel a long way in this time. The chart shows the stopping distances at different speeds.

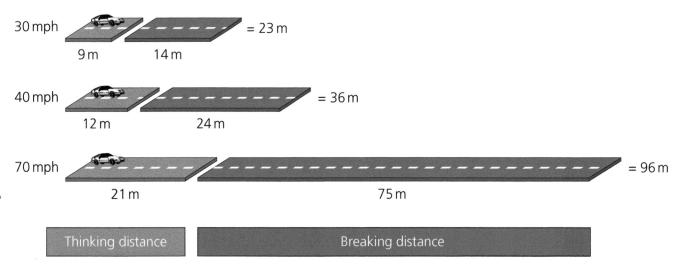

30 mph   9 m   14 m   = 23 m

40 mph   12 m   24 m   = 36 m

70 mph   21 m   75 m   = 96 m

Thinking distance   Breaking distance

▲ **Stopping distances at different speeds**

7 If you can react to danger in a second, how long is the total stopping distance if you are travelling at:

   a 30 mph

   b 40 mph?

8 a How long is it from the front to the back of your classroom?

   b What is the highest speed you could go at and still stop between the front and back of your classroom?

9 Racing car drivers have very fast reaction times. Suggest two other types of people who need fast reactions. Give reasons for your choices.

Changes to stopping distances

Many things can make the reaction time longer:

* drinking alcohol

* some types of drugs

* being distracted, perhaps by a phone, someone talking, or listening to the radio

* being very tired.

The braking distance can also get longer if the:

* road is icy or wet

* brakes or tyres have not been maintained properly.

10 If alcohol doubles your reaction time to 2 seconds, how will it affect the total stopping distance at 40 mph?

11 a Will icy conditions affect your reaction time?

b If icy roads made the braking distance twice as long, how long would the total stopping distance be at 30 mph?

Activity

Testing your reaction time
Try at least two methods to test your reaction time. You could use a computer testing program or a dropping stick test.

For the dropping stick test:

1 In pairs decide who will have their reaction time tested and who will be the tester.

2 The tester should hold two 30 cm rulers vertically at the end near the 30 cm mark, one in each hand.

3 The other person places their index finger and thumb of each hand either side of the 0 cm marks, holding them as wide as possible ready to catch a ruler when one falls.

4 Without warning, the tester lets go of one of the rulers and the other student tries catch the ruler as soon as possible.

5 Write down the distance travelled by measuring where the ruler was caught just above the student's first finger.

6 Do this again two more times, making sure that the ruler is always dropped at random times.

Do they give you the same result?

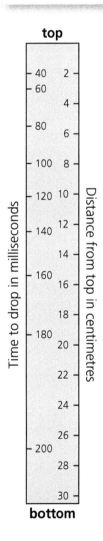

Summary

✳ Forces

» A force is a push or pull on an object by another object.

» When the objects are touching, the forces are contact forces. Contact forces include pushes and pulls, friction, air resistance and tension.

» When objects are not touching, the forces are called non-contact forces. Non-contact forces include electrostatic attraction or repulsion, magnetic attraction and repulsion and gravity.

» When a force moves an object, the force has done work.

✳ Speed and safety

» Speed is the distance travelled divided by the time taken. The usual units for laboratory work are metres per second. Road speeds tend to be measured in miles per hour.

» Stopping distance for cars depends on thinking time and braking time.

» Alcoholic drink, some drugs, distractions and tiredness all increase the time for the driver to react. This raises the thinking time but not the braking time.

» Icy or wet roads, poor tyres or brakes can increase the braking time but should not affect thinking time.

» The faster the car is travelling, the more damage is done in a collision.

Show you can...

A Discuss the differences between renewable and non-renewable energy resources. List the energy sources you know about into renewable and non-renewable categories.

B Prepare a set of posters for a car showroom. The posters should show the safety features in their cars. Make sure you mention air bags, seat belts, the steel cage for the passengers and crumple zones. Include labels which give the scientific reasons why these features improve safety for car occupants in a crash.

C Create a presentation of at least five slides showing a variety of forces in action. For each force, include a short pop-up caption to explain what is happening.

D Prepare a leaflet to hand out at school campaigning for a 20 mph speed limit on the road outside.

E Build and test a spring-powered forcemeter. Explain how this could be used to test the weight of some objects.

Episode 1 Stage show

✳ Wiring it up

▲ **Arcadia are a music group that put on gigs using a giant metal spider**

It needs four articulated lorries to travel in and runs on biodiesel. A crew of 15 people take two days to set it up. It has hundreds of metres of electrical circuits and hundreds of lights. It's the biggest steel spider in Bristol!

The core of the Arcadia spider's electric cables is copper. Copper is a good conductor of electricity. The outer layer will be plastic. Plastic is an insulator so electricity cannot pass through it.

Electric current

An electric current is a flow of charge along a conductor. The more charge that flows every second, the bigger the current. The charge carries energy along the wire. This energy makes the lights and speakers on the spider work.

Charge flows because there is a difference in the amount of charge between the two ends of the wire. This difference is measured in volts. The bigger the difference (voltage), the more energy that can be carried. So the current is bigger.

1 What was the best gig you have ever been to? Why was it so good?

2 List all the devices on the Arcadia spider that need electricity to work.

Practical

Circuit testing
Build a device to test if a circuit is complete. How might this be useful to the engineers who fit the electrical cabling in place on the Arcadia spider?

3 List:

   a two electrical conductors

   b two insulators.

4 Why are electrical wires always covered in an insulating layer?

5 Give one way to increase the current flowing in a conductor.

6 Where does the energy to light the lights come from on the Arcadia spider?

Resistance

Not all conductors are the same. Conductors that let electric current flow very easily have a low resistance. Conductors that do not let the current flow easily have a high resistance.

Radio microphones contain a small radio transmitter. This sends the signal to the sound system. This means no wires so people can move around more easily. They need a battery to power the radio. Scientists call these batteries cells.

▲ **This performer is holding a radio microphone (mic)**

Practical

Battery power
Measure the output from some simple cells. Which ones produce the greatest current?

To understand the difference between charge and current think about water being carried from a water store to a plant. The water store is held above the ground. The higher it is the faster the water flows along the pipe. The height is a model of the potential difference.

The pipe could be narrow or wide. The wider the pipe the more water flows. The flow of water is the electric current.

A narrow pipe only lets a little water pass. The size of the pipe is like the resistance of the wire in an electric circuit.

▼ **A pipe as a charge carrier**

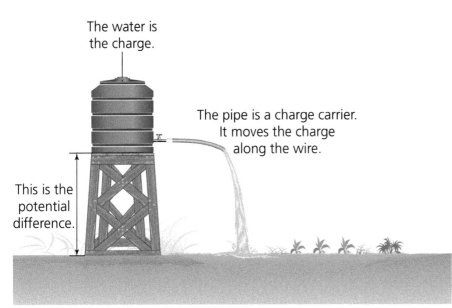

The water is the charge.

The pipe is a charge carrier. It moves the charge along the wire.

This is the potential difference.

7 Give one advantage of a radio mic compared with a wired microphone.

▲ **Scientists and electricians call these cells, not batteries**

Practical

Good conductor?
Carry out some tests on a variety of different wires to see which has the lowest resistance.

8 a What is electrical resistance?

b Should a good electric cable have a high or low resistance? Why?

Episode 2 Power at home

✳ The world's greatest plug

▲ Is this a safe way to plug in all your devices?

When a battery produces a current, it flows in the same direction at a constant rate until the battery runs out. It is called direct current or d.c. Batteries typically produce a voltage of 1.5 volts.

The mains electricity in buildings uses alternating current or a.c. In a.c. the direction of flow changes regularly. In the UK it changes about 50 times a second! a.c. is used instead of d.c. for the mains because it is easier to move electricity around the country by a.c. than by d.c.

▼ Alternating current (a.c.) and Direct current (d.c.)

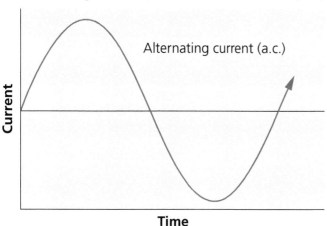

Alternating current (a.c.)

Current

Time

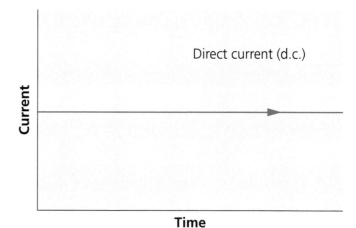

Direct current (d.c.)

Current

Time

▼ **Mains electricity in a house showing circuit board and outlets**

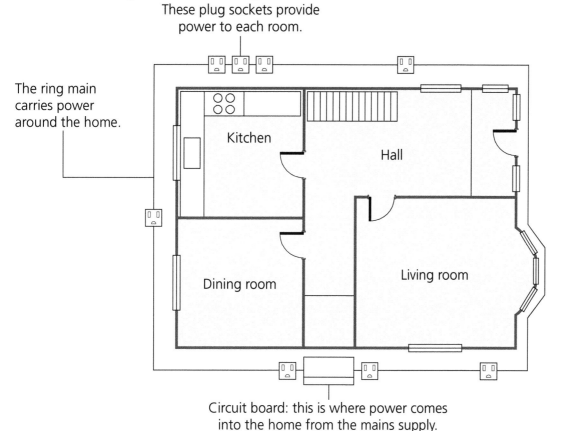

These plug sockets provide power to each room.

The ring main carries power around the home.

Circuit board: this is where power comes into the home from the mains supply.

The circuit board is the place where the electricity from the power station enters your house.

The ring main is a set of cables that take power to each room in your house.

A socket is a safe way to get electrical energy from your ring main and into your appliances.

The plug is the way to connect an appliance to the power in the ring main.

1 a Count the number of sockets in each of your rooms at home.

  b Which room has the most sockets? Why do you think this is?

2 Give two differences between electricity from a battery and electricity from the mains.

3 a Where is the circuit board at your home?

  b Why might an electrician switch off all the power to your home before working on a power socket?

Plugs and wires

Larger devices like electric cookers, washing machines and large televisions need a lot of power. So the current flowing in the wires is large as well. This makes the electricity more dangerous than the low voltages and small currents that come out of little batteries.

Most electrical devices are connected to mains power using a 3-pin plug.

▼ **A 3-pin plug**

The blue wire is the neutral wire. This takes the current back to the circuit board.

The large earth pin connects to the green and yellow striped earth wire.

The live wire carries the current. It is coloured brown. It goes to the left when you look at an open plug.

The fuse is a safety device. If too much current flows through the fuse it will break and the current stops.

▲ **This metal kettle is potentially dangerous**

If the live wire touches the metal case of a kettle, that kettle will become live. This means that electricity is flowing through it. If you touch it, the electricity will flow through you down to the ground and then back to the power station. Over 200 000 people were seriously injured by an electric shock in the UK in 2014!

The earth wire is a way to make the kettle safe. It is attached to the metal kettle body. It conducts any power in the metal case back to the power station so it does not go through you.

A radio often has a plastic case. This will not conduct electricity. The devise is double insulated – once by the plastic coatings on the wires, and once by the plastic case. These often do not have an earth wire.

4 a Why is the body of a plug made of plastic, not metal?

b Why should you never unscrew a socket in the wall when the power is on?

5 A hairdryer has a plastic case. Why does it not need an earth wire?

Fuses

Fuses are weak links in a circuit. They are made of thin metal. If the circuit suddenly starts to carry too much current, the wire melts. The electricity stops flowing. We say the fuse has blown. It is much better to replace one blown fuse than to replace a whole television!

But a fuse that is too weak means it keeps blowing and has to be replaced often. A fuse that is too strong may not blow to protect the device.

The power used by an appliance is measured in watts. It is usually given on a small label near the back of the device. Imagine a device runs on 230 watts. How much current would it take to deliver the 230 watts of power?

power = voltage × current flowing in amps

$P = V \times I$

230 W = 230 V × the current in amps

Current = 230/230 = 1 amp

So a fuse for 3 amps would be perfect. The television can get all the power it needs with a current of 1 amp. If the current suddenly went above 3 A, then the fuse would blow before the television was damaged.

Paying for electricity

An electricity bill tells you how much energy you have used. The total energy depends on:

* the power delivered to each device

* how long the device was switched on for.

If you watch a whole episode of your favourite show your television will be on for 1 hour. If it needs 230 watts per second, we need to multiply 230 by the number of seconds in an hour:

total power needed = 230 × 60 × 60 = 828 000 watt-hours

To make the numbers easier we can divide by 1000 to give kilowatt-hours.

828 000 watt-hours = 828 kWh

Your electricity bill charges you for the kiloWatt-hours of power you use.

▲ The fuse has very thin wire that will melt if too much current flows through it

▲ The colour of the fuse case tells you how much power it can let pass

6 What size fuses would be good for each of these devices?

a A laptop computer using 100 W

b An electric cooker that needs 3000 W when all the rings and oven are working

c A 50-inch plasma TV that needs 450 W?

Episode 3 Magnets

✱ Magnets in action

▲ How are these fruits stuck on the fridge door? It's not glue. Each fruit has a small magnet in it. The magnet sticks to the steel door of the fridge

1 How many magnets are there in your home? List as many as you can.

2 Do magnets attract all metals? How could you find out?

3 A fridge seal also has a magnet in it. How does this help the fridge to work better?

Practical

Test your pulling power
The plastic letters are very light. If they were too heavy, the magnets could not stick them to the fridge door. Plan a test to measure the strength of a set of magnets.

Poles apart

Magnets can attract metals like steel. You can feel this magnetic force when you pull a magnet from a fridge door. The parts of the magnet where the force is strongest are called the poles. In a simple bar magnet, the poles are usually at opposite ends. Every magnet has two poles:

✱ the north-seeking pole (N)

✱ the south-seeking pole (S).

If you push two North poles together they will repel each other. Two South poles also repel each other. Only when the poles are different do the magnets attract each other strongly. This is a non-contact force.

▼ Magnetic attraction and repulsion

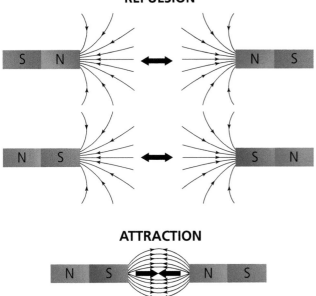

Magnetic fields

Around the poles of a magnet is an invisible magnetic field. We know it is there because we can feel the force acting on another magnet. It is possible to map the magnetic field. The lines are called magnetic field lines. The closer they are together, the stronger the field.

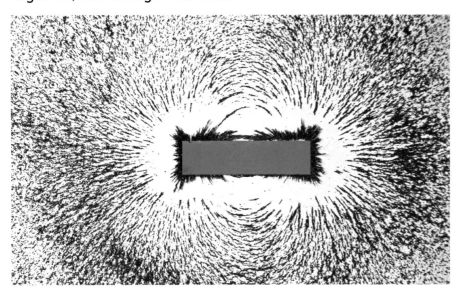

▲ A bar magnet and its magnetic field

The tractor beam!

▲ **You have made your escape from the enemy's ship, but a magnetic tractor beam manages to drag you back!**

The magnetic tractor beam only exists in films! But we can make a magnet that can be switched off and on, like the tractor beam.

When an electric current flows through a wire, the current creates a magnetic field around it. The field spreads out from the wire. If the wire is made into a coil called a solenoid, the magnetic field becomes stronger. However, when the current is switched off, the magnetic field disappears.

▼ **A coil of wire and its magnetic field lines**

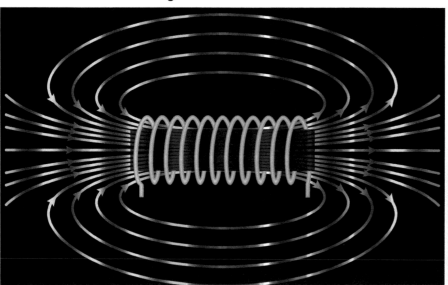

4 a What is a solenoid?

b What does the field from a solenoid look like?

Activity

Pulling up
Investigate the factors that affect the strength of an electromagnet. How much could you lift with your electromagnet?

Using electromagnets

A magnet that can be switched off and on is useful. It can sort iron scrap from a pile of mixed metals. It can also pick out steel cans from waste metals for recycling, and leave the non-magnetic aluminium cans behind.

▲ **Electromagnets are used to sort scrap metals**

Relays

A relay is a way to control a circuit with an electromagnet. It uses a small electromagnet and a springy steel strip.

When current flows, the electromagnet pulls on a springy steel strip. This completes the other circuit.

When the current in the coil stops, the electromagnet switches off. The steel strip springs back and breaks the connection for the other circuit. In this way, relays can operate switches from a long way away.

▲ **Relays operate switches at a distance**

Summary

✳ Charge and current

» An electric current is a flow of electrical charge along a conductor. The more charge that flows, the higher the current.

» The current through a circuit depends on the resistance of the circuit and the voltage across it. The greater the resistance, the smaller the current. The greater the voltage, the greater the current.

» Electrical current is measured in amps using an ammeter.

» Voltage is measured in volts using a voltmeter.

✳ Magnets and electromagnets

» When two magnets are brought close together, they exert a force on each other. Two like poles repel each other. Two unlike poles attract each other.

» The poles of a magnet are the places where the magnetic forces are strongest. Attraction and repulsion between two magnetic poles are examples of non-contact forces.

» Magnetic fields can be shown as a diagram with magnetic field lines showing the direction of the magnetic field.

» When a current flows through a conducting wire, it produces a magnetic field. The strength of the magnetic field at a point depends on the size of the current passing along the wire and the distance from the wire.

» A solenoid is a coil of conducting wire. This shape increases the strength of the magnetic field produced when a current flows. Adding an iron core also increases the magnetic field strength of a solenoid. An electromagnet is a solenoid with an iron core.

» The magnetic field of an electromagnet disappears when the current is switched off.

» Electromagnets are used for sorting metal scrap and in relays to control other devices.

✳ Electricity at home

» Cells and batteries supply current that always passes in the same direction. This is called direct current (d.c.). An alternating current (a.c.) is one that changes direction. Mains electricity is an a.c. supply. In the UK it has a frequency of 50 Hz and is 230 V.

» Most electrical appliances are connected to the mains using a three-core flex. The insulation covering each wire in the flex is colour-coded. Live wire: brown; neutral wire: blue; earth wire: green and yellow stripes.

» The earth wire can conduct any electricity safely if a metal casing becomes live. Some appliances do not have an earth wire because they are double insulated or have plastic cases.

» A fuse contains a thin piece of wire that melts if the current gets too large. This cuts the circuit to prevent damage to the appliance. The correct size for the fuse depends on the power needed by the appliance.

» Everyday electrical appliances are designed to bring about energy transfers. The amount of energy an appliance transfers depends on how long the appliance is switched on for and the power of the appliance.
energy (kWh) = power (kW) × time (h)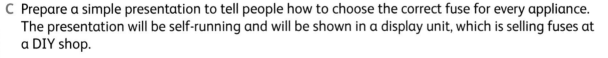

Show you can...

A Have a discussion with a partner about electricity. What is it? What can it do? What are the dangers of electricity and how can we keep ourselves safe when using it? Use the words current, voltage, conductor and insulator in your discussion.

B Produce a poster showing as many different uses of electromagnets as you can.

C Prepare a simple presentation to tell people how to choose the correct fuse for every appliance. The presentation will be self-running and will be shown in a display unit, which is selling fuses at a DIY shop.

D Survey your home to find out where all the mains sockets are placed. Draw a plan showing the layout and the sockets. Identify rooms where you might need more sockets and some where you do not use all of the ones you have. Explain these differences in power use in each room.

E Build and test a model electromagnetic door lock. In the real world it should be possible to lock or unlock the door from the other side of the room.

Episode 4 Waves and wobbles

✳ Sound it out

1 Would you like to learn to play a musical instrument? Why? Why not?

2 List all the different ways to make a sound you know. For example, hitting a bell.

Activity

Sound it out
Build a simple sound box with one string. Find a way to change the pitch and volume of the note the string makes when it is plucked. Try to get it to play a tune.

▲ A simple sound box

▲ The sound from the bass guitar travels through the air to the ears of the people in the audience

Describing waves

Waves are to-and-fro movements that travel through a medium. A medium is the thing that carries the wave.

◀ For these surfers the water is the medium. They plan to ride the wave all the way to the beach

The wave travels a long way, but the water moves very little. When a wave passes a surfer paddling out from the beach, the surfer moves up to the crest and then down to the trough of the wave. The distance from the lowest or the highest point of a wave to the middle is called the amplitude.

3 a What is the amplitude of a wave?

   b Some giant waves travel hundreds of kilometres across the ocean. They may be 50 metres from top to bottom. How far does the water travel in this wave?

The surfer starts at the top, or crest, of the passing wave.

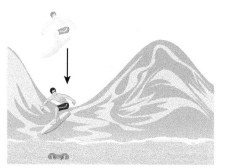

As the wave passes the surfer moves down to the trough of the wave.

When the next wave comes, the surfer moves up again. The wave moves a long way onto the beach but the surfer moves only a short distance up and down.

▲ **How a surfer moves**

Surfers look for waves with a large amplitude. They also need to know how often waves pass. The number of waves that pass a point in a second is called the frequency. A beach with high amplitude waves and a good frequency is surfer paradise!

The distance between a point on a wave and the same point on the next one is called the wavelength.

Mobile phones use radio waves that have a wavelength of about 15 cm. When the crest of a radio wave hits your phone, the next one is still 15 cm away. The wavelength of a sound wave is longer. The note made by the middle string on a guitar has a wavelength of about 60 cm.

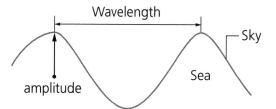

▲ **Total distance moved by water or surfer is 1 metre**

4 a Why do surfers like waves with a large amplitude?

   b Draw a diagram of a wave. Add labels to show the crest, trough, wavelength and amplitude.

Types of waves

Waves in water are transverse waves. They vibrate from side to side. Sound waves are different.

Sound uses air as the medium. When a guitar string vibrates it pushes on the air around it. It creates areas of crowding of air molecules and areas of spreading out of air molecules in the air around it. These waves are called longitudinal waves because the vibration runs along the direction of the wave.

▼ Particles move close together or further apart in a longitudinal wave

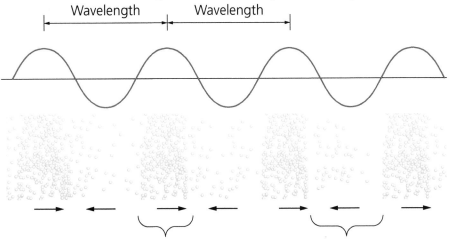

Wave speed

Sound waves travel at about 1200 km/h. It would take a sound wave about 18 minutes to get from Edinburgh to Leeds. Light waves travel much faster. A light wave would do the same journey in under 1 second!

This equation shows the link between frequency, wavelength and wave speed.

wave speed (m/s) = frequency (Hz) × wavelength (m)

5 a A note from a guitar has a frequency of 30 Hz. Its wavelength is 1.3 m. How fast is the sound travelling?

b How long will the note take to get to the back of the hall? The back row is 60 metres from the stage.

Episode 5 Death rays and diagnosis

✱ Cook that pigeon!

It seemed like a good idea at the time. Build a large sculpture shaped like a bowl. Give it a silver surface.

Unfortunately, at some times in the summer, the mirror focuses the reflected sunlight. It goes back into the air as a beam of light that could cook passing birds!

▲ This is the Sky Mirror by Anish Kapoor

Practical

Boiling the kettle
In Tibet people use sunlight to boil water. Giant mirrors focus the sunlight onto the bottom of a kettle. Build and test a simple solar kettle. What things make it work well?

Rainbows

Sunlight is an electromagnetic wave. Electromagnetic waves are transverse waves. It contains a mix of wavelengths. Visible light is a small part of the electromagnetic spectrum. A prism can sort the different wavelengths of light out to make a rainbow.

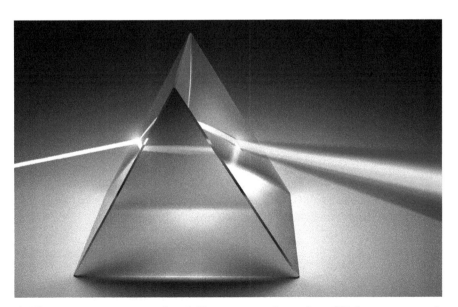

▲ This prism is sorting the different wavelengths of light out to make a rainbow. The red end has the longest wavelength, the blue end has the shortest wavelength

1 Do you think the Sky Mirror would really cook birds flying overhead? Why? Why not?

2 How could you protect the birds just in case?

3 What does the word 'visible' mean?

4 What colours of light have a longer wavelength than yellow light?

The full spectrum

All electromagnetic waves travel at the same speed in a vacuum. The speed is roughly 300 000 000 metres every second! A flash of light gets from Edinburgh to London in under 1 second!

There are different types of electromagnetic radiation. The differences depend on the wavelength or frequency of the wave. The speed always remains the same. The cards below shows these different types of radiation.

5 a Which type of radiation has more energy than X-rays?

b Which type of radiation has the longest wavelength?

Frequency goes up, wavelength gets shorter

Frequency goes down, wavelength gets bigger

Gamma ray

Gamma rays are very high energy radiation. These can cause illness very quickly.

X-ray

X-rays contain a lot of energy, and can pass straight through the human body!

Ultraviolet

Ultraviolet light in sunshine gives a nice tan, but can also cause sunburn.

Visible light

We can see visible light-without it we would be blind! The fibre optic cables that carry telephone, television and the internet into your home also use visible light to carry the information Fibre optic.

Infrared

Infrared radiation is invisible but we can often feel it as heat. Infrared gears work this way. Infrared is also used in some remote controls.

Microwaves

Microwave radiation heats up water as it passes through it. So they are used in microwave ovens to heat food from the inside. Satellites also use microwaves to carry information.

Radio waves

Radio waves carry signals to our radios, televisions and mobile phones. Bluetooth speakers often use low power radio waves to carry music from your phone without needing a cable.

▲ **The electromagnetic spectrum**

Radiation danger

Ultraviolet light in sunshine gives a nice tan. But it can also cause sunburn. The radiation delivers energy and damages cells in the skin. Sometimes the damage is so serious it causes skin cancer, maybe many years later.

The radiation used for X-rays contains a lot of energy. It can pass straight through your body! X-rays can cause cancers deep in the body. Pregnant women need to be very careful because X-rays can cause problems for the growing foetus.

Gamma rays are very high energy radiation. These can cause illness very quickly.

The damage done by radiation depends on the type of radiation and the size of the dose. A large dose in a short time is very dangerous, but even a low dose over a long time can cause problems.

Nuclear radiation

Some atoms are unstable. This means the atoms split apart and give out radiation. There are three sorts of nuclear radiation: alpha, beta and gamma.

Radiation can be useful. Alpha radiation is used in smoke detectors. Some fruits are exposed to low doses of beta or gamma radiation to keep them fresh for longer. Gamma radiation can also be used to treat some cancers - the radiation destroys the cancer deep in the body.

▲ Some of these people could end up with sunburn or even cancer

▲ X-rays are used to see inside the body

6 Why do doctors say you should not have too many X-rays in a year?

▲ Gamma rays are given out by nuclear explosions

Alpha particles are parts of an atomic nucleus that shoot out with a positive charge. They are slow moving and can be blocked by a piece of paper, your skin or even air.

Beta particles are small, negatively-charged, high-speed particles. They can pass through paper or your hand but are stopped by a thin layer of plastic or aluminium.

Gamma radiation is part of the electromagnetic spectrum and travels at the speed of light. It can only be blocked by thick layers of lead or concrete.

Hand   **Aluminium**   **Concrete**

Summary

Describing waves

» Waves are to-and-fro movements in a medium. They pass through the medium.

» In a transverse wave, the oscillations are at right angles to the direction of energy transfer. The ripples on a water surface are an example of a transverse wave.

» In a longitudinal wave, the oscillations are parallel to the direction of energy transfer. Longitudinal waves show areas of compression and rarefaction. Sound waves are longitudinal.

» The amplitude of a wave is the distance between the middle of the wave and its highest, or lowest, point.

» The wavelength of a wave is the distance from a point on one wave to the equivalent point on the adjacent wave.

» The frequency of a wave is the number of waves passing a point each second.

» You may need to use the equation:
wave speed (m/s) = frequency (Hz) × wavelength (m)

Using waves

» Electromagnetic waves are transverse waves that transfer energy.

» All types of electromagnetic wave travel at the same velocity through a vacuum (space) or air. The waves that form the electromagnetic spectrum are grouped in terms of their wavelength and their frequency.

» These groups are: (longer wavelengths) radio, microwave, infrared, visible light (red to violet), ultraviolet, X-rays and gamma-rays. (shorter wavelengths).

» Electromagnetic waves with shorter wavelengths tend to carry more energy. Ultraviolet waves, X-rays and gamma rays can damage human cells. The amount of damage depends on the type of radiation, the level of the exposure and how long the person was exposed to the radiation.

Show you can...

A Explain to a friend why it is not a good idea to have an X-ray more often than you need to.

B Produce a poster showing the different types of electromagnetic radiation, their uses and possible dangers.

C Prepare an animation of a wave pattern to show how the parts of the medium move a very short distance but the wave itself can travel a long way. Include a slide showing the wavelength and amplitude marked on a wave shape.

D Describe how the solar kettles used in places like Tibet work. Why are they particularly useful in Tibet?

E Build and test a stringed instrument that can play a recognisable tune. Explain how the notes can change pitch and volume.

Specification coverage*

AQA Entry Level Certificate in Science Specification Content * For the latest mark schemes, please also refer to the AQA website.	Combined Science Specification references		Hodder ELC Science textbook	
	Trilogy	Synergy	Section/ Episode	Topic
Component 1 – Biology: The human body				
3.1.1 **What is the body made of?** Cells are the basic building blocks of all living organisms.	4.2.1	4.1.3.2	Biology 1 1: Inside the Body	Vampires and monsters
Most human cells are like most other animal cells and have the following parts: 1 nucleus – controls the activities of the cells and contains the genetic material 2 cytoplasm – where most chemical reactions take place 3 cell membrane – controls the passage of substances in and out of cells. Cells may be specialised to carry out a particular function, e.g. sperm cells, nerve cells and muscle cells. Students should be able, when provided with appropriate information, to explain how the structure of different types of cell relates to their function.	4.1.1.3	4.1.3.2		
A tissue is a group of cells with a similar structure and function. Students should develop some understanding of size and scale in relation to cells, tissues, organs and systems.	4.2.1	4.2.1.2		
Organs are aggregations of tissues performing similar functions. Organs are organised into organ systems which work together. Students should be able to identify the position of the major organs in the human body such as the brain, heart, liver, lungs, kidneys and reproductive organs. Students should be able to identify the function of the main organ systems.	4.2.1	4.2.1.2		

		The human circulatory system consists of the heart, which pumps blood around the body (in a dual circulatory system) and blood, which transports oxygen, proteins and other chemical substances around the body. Students should be able to recognise the different types of blood cell from a photograph or diagram.	4.2.2.2	4.2.1.3	Biology 1 3: Going faster	Pump it up!
		The human digestive system contains a variety of organs: • salivary glands • stomach • liver • gall bladder • pancreas • small intestine • large intestine. Students should be able to identify the position of these organs on a diagram of the digestive system. Enzymes are used to convert food into soluble substances that can be absorbed into the bloodstream.	4.2.2.1	4.2.1.5	Biology 1 2: Britain's big problem	Break it down
3.1.2	How the body works	Respiration releases the energy needed for living processes and is represented by the equation: glucose + oxygen → carbon dioxide + water (+ energy) Students should know the word equation for respiration. Students should know that glucose is derived from the diet and that carbon dioxide and oxygen gases are exchanged through the lungs.	4.4.2.1	4.2.1.1	Biology 1 3: Going faster	Muscles and marathons
		Lifestyle can have an effect on people's health e.g. diet and exercise are linked to obesity; smoking to cancer; alcohol to liver and brain function. A healthy diet contains the right balance of the different food groups you need and the right amount of energy.	4.2.2.5 and 4.2.2.6	4.3.1.2	Biology 1 2: Britain's big problem	Let's get fitter
		People who exercise regularly are usually fitter than people who take little exercise.	4.2.2.6	4.3.1.2		
3.1.3	How the body fights disease	Infectious (communicable) diseases are caused by microorganisms called pathogens.	4.3.1.1	4.3.3.1		

		These may reproduce rapidly inside the body and may produce poisons (toxins) that make us feel ill. Viruses damage cells in which they reproduce.	4.3.1.1	4.3.3.1		
		White blood cells help to defend against bacteria by ingesting them.	4.3.1.6	4.3.3.4		
		Vaccination involves introducing small quantities of dead or inactive forms of a pathogen into the body to stimulate the white blood cells to produce antibodies so that if the same pathogen re-enters the body, antibodies can be produced rapidly. Students should be able to explain the use of vaccination in the prevention of disease.	4.3.1.7	4.3.3.5	Biology 1 4: Diseases	Attack!
		Medical drugs are developed and tested before being used to relieve illness or disease. Drugs change the chemical processes in people's bodies. People may become dependent or addicted to the drugs and suffer withdrawal symptoms without them.	4.3.1.9	4.3.3.7		
		Antibiotics, including penicillin, are medicines that help to cure bacterial disease by killing infective bacteria inside the body, but cannot be used to kill viruses.	4.3.1.8	4.3.3.6		
3.1.4	How the body is coordinated	The human body has automatic control systems, which may involve nervous responses or chemical responses coordinated by hormones.	4.5.2	4.3.1.4	Biology 1 5: Controlling the body	Danger!
		Reflex actions are automatic and rapid. Examples include the response of the pupil in the eyes to bright light, and the knee jerk reaction.	4.5.2	4.2.1.6		
		Hormones are secreted by glands and are transported to their target organs by the bloodstream.	4.5.3.1	4.2.1.7		
		Several hormones are involved in the menstrual cycle of a woman, including some that are involved in promoting the release of an egg. Students should be familiar with a diagram of the menstrual cycle.	4.5.3.3	4.3.1.6		

		The uses of hormones in controlling fertility include: • giving oral contraceptives that contain hormones to inhibit eggs from maturing • giving 'fertility drugs' to stimulate eggs to mature.	4.5.3.4 and 4.5.3.5	4.3.1.7 and 4.3.1.8	Biology 1 5: Controlling the body	Danger!
		Students should be able to evaluate the benefits of, and the problems that may arise from, the use of hormones to control fertility.	4.5.3.4 and 4.5.3.5	4.3.1.8		
3.2.1	**What are the feeding relationships between living organisms?**	Radiation from the Sun is the source of energy for living organisms.	4.4.1.1	4.2.2.5	Biology 2 1: Up the garden path	Welcome to Rooftops!
		Green plants and algae absorb a small amount of the light that reaches them and make glucose by photosynthesis. These organisms are called producers. Carbon dioxide + water → glucose + oxygen Students should know the word equation for photosynthesis.	4.4.1.1	4.2.2.5		
		Animals and plants may be adapted for survival in the conditions where they normally live.	4.7.1.4		Biology 2 1: Up the garden path AND Biology 2 5: Pets and planets	Welcome to Rooftops! AND A suitable pet
		Feeding relationships within a community can be represented by a food chain. All food chains begin with a producer. A food web can be used to understand the interdependence of species within an ecosystem in terms of food resources.	4.7.2.1	4.4.2.1	Biology 2 1: Up the garden path	Food chains
		All materials in the living world are recycled to provide the building blocks for future organisms.	4.7.2.2	4.4.1.2	Biology 2 2: Death and decay	Cheating death?
		Decay of dead plants and animals by microorganisms returns carbon to the atmosphere as carbon dioxide to be used by plants in photosynthesis.	4.7.2.2	4.4.1.2		

3.2.2	What determines where particular species live?	Plants often compete with each other for light and space, and for water and nutrients from the soil.	4.7.1.1	4.4.2.2	Biology 2 1: Up the garden path AND Biology 2 5: Pets and planets	Welcome to Rooftops! AND A suitable pet
		Animals often compete with each other for food, mates and territory.	4.7.1.1	4.4.2.2		
		Animals and plants are subjected to environmental changes. Such changes may be caused by non-living or living factors.	4.7.1.2 and 4.7.1.3	4.4.2.3		
		Pollution of the environment can occur: • in water, from sewage, fertiliser or toxic chemicals • in air, from smoke and gases such as sulfur dioxide which contributes to acid rain • on land, from landfill and from toxic chemicals such as pesticides and herbicides, which may be washed from land into water. Students should recognise that rapid growth in human population means that more resources are used and more waste is produced.	4.7.3.2	4.4.2.6	Biology 2 3: Safe Earth	Let's save the world!
3.2.3	How life has developed on Earth	Darwin's theory of evolution states that all species of living things have evolved from simple life forms that first developed more than three billion years ago.	4.6.2.2	4.4.4.2	Biology 2 5: Pets and plants	Life story
		In natural selection, individuals with characteristics most suited to their environment are most likely to survive to breed successfully.	4.6.2.2	4.4.4.2		
		Artificial selection (selective breeding) is the process by which humans breed plants and animals for particular genetic traits.	4.6.2.3	4.4.4.5		
		There are two types of reproduction: • sexual reproduction, which involves the joining of male and female sex cells. There is a mixing of genetic information, which leads to variety in the offspring • asexual reproduction, where only one individual is needed as a parent. There is no mixing of genetic information, which leads to identical offspring (clones).	4.6.1.1	4.4.3.1		

		The genetic material in the nucleus of a cell is made of a chemical called DNA, which is contained in structures called chromosomes. Students should know that a cell consists of a nucleus that controls the actions of the cell, and cytoplasm.	4.6.1.3	4.4.3.1	Biology 2 4: Sex and survival	Animal attraction
		Chromosomes carry genes that control the characteristics of the body. Humans have 23 pairs of chromosomes. Only one pair carries the genes that determine sex: females have the same sex chromosomes (XX); in males the chromosomes are different (XY).	4.6.1.6	4.4.3.1		
		In genetic engineering, genes from chromosomes of humans and other organisms can be 'cut out' and transferred to the cells of other organisms. Students should be aware of the potential benefits and risks of genetic engineering.	4.6.2.4	4.4.4.6		
3.3.1	Atoms, elements and compounds	All substances are made of atoms. An atom is the smallest part of an element that can exist.	5.1.1.1		Chemistry 1 3: Magicians and makers	The first artists?
		A substance that is made of only one sort of atom is called an element. There are about 100 different elements. Elements are shown in the periodic table. Metals are towards the left and the bottom of the periodic table and non-metals towards the right and the top of the periodic table. Students should know that most of the elements are metals.	5.1.1.1 and 5.1.2.3	4.5.1.1 and 4.5.1.2	Chemistry 1 3: Magicians and makers AND 4: Sorted!	The first artists? AND Metal or non-metal? AND Get into groups!
		Elements in the same group of the periodic table have similar chemical properties.	5.1.2.1	4.5.1.1	Chemistry 1 3: Magicians and makers AND 4: Sorted!	The first artists? AND Get into groups!

		When elements react, their atoms join with other atoms to form compounds.	5.1.1.1	4.5.2.1	Chemistry 1 3: Magicians and makers	The first artists?
		Some compounds are made from metals combined with non-metals, for example sodium chloride and magnesium oxide. Students should be able to recognise simple compounds from their names, e.g. sodium chloride, magnesium oxide, carbon dioxide. Some compounds are made from only non-metals, for example carbon dioxide.	5.2.1.2	4.6.2.1		What's in a name?
		Chemical reactions can be represented by word equations. Students should be able to write word equations for reactions of metals and non-metals, reactions of non-metals to produce oxides, and the other chemical reactions in this specification.	5.1.1.1	4.5.2.1		
3.3.2	How structure affects properties	The three states of matter are solid, liquid and gas. Melting and freezing take place at the melting point, boiling and condensing take place at the boiling point. The three states of matter can be represented by a simple model.	5.2.2.1	4.1.1.1	Chemistry 1 2: Melting and evaporating	Liquid chocolate
		When a solid melts to become a liquid the particles are able to move about but stay close together. When a liquid boils and becomes a gas the particles separate and move about rapidly. Substances with high melting points have strong forces that hold their particles together. Substances with low boiling points have weak forces between their particles.	5.2.2.1	4.6.2.5		
		Diamond and graphite are forms of the element carbon with different properties because of their different structures. Diamond is hard because the carbon atoms are joined together in a giant three dimensional structure. Graphite is slippery because the carbon atoms are joined together in layers that can slide over each other.	5.2.3.1 and 5.2.3.2	4.8.1.1	Chemistry 1 5: Metal treasures	Carbon

3.3.3	Separating mixtures	A mixture contains two or more substances not chemically combined together. Mixtures can be separated by processes such as filtration, distillation, crystallisation and chromatography.	5.1.1.2	4.2.2.4	Chemistry 1 1: Substances	The pure stuff
		Paper chromatography can be used to separate mixtures and can give information to help identify substances. In paper chromatography a solvent moves through the paper carrying different compounds different distances.	5.8.1.3	4.2.2.4		
3.3.4	Metals and alloys	Unreactive metals, such as gold, are found in the Earth as the metal itself, but most metals are found as compounds that require chemical reactions to extract the metal. Metals less reactive than carbon can be produced by heating the metal compounds in the ore with carbon. Ores contain enough metal to make it economic to extract the metal. Large amounts of rock need to be quarried or mined to get metal ores.	5.4.1.3	4.8.2.1	Chemistry 1 5: Metal treasures	Ancient metals
		We should recycle metals to save resources and limit environmental impacts. Students should be able to describe the social, economic and environmental impacts of mining ores and recycling metals.	5.10.2.2	4.8.2.9		Recycling metal
		Metals have giant structures of atoms with strong bonds between the atoms and so most metals have high melting points.	5.2.2.7	4.6.2.7		
		Metals are good conductors of electricity and thermal energy. Copper has properties that make it useful for electrical wiring and plumbing.	5.2.2.8	4.6.2.7		
		Aluminium is a useful metal because of its low density and resistance to corrosion.				

		Most metals in everyday use are alloys. Pure iron, gold and aluminium are too soft for many uses and so are mixed with small amounts of other elements to make alloys, which are harder for everyday use.	5.2.2.7	4.6.2.7		
		Most iron is converted into steels. Steels are alloys since they are mixtures of iron with carbon and other metals.				
3.3.5	**Polymers**	Polymers such as poly(ethene), poly(propene) polystyrene and PVC are made from small compounds called monomers that join together to form very long chains. Polymers are waterproof, resistant to chemicals, and can be moulded, so they have many useful applications as packaging materials, pipes and containers.	5.2.2.5	4.6.2.4	Chemistry 1 6: Plastics	Plastics
		Many polymers are not biodegradable, so they are not broken down by microbes. This can lead to problems with waste disposal.				
3.4.1	**Reactions of acids**	Acids react with some metals to produce salts and hydrogen. Hydrochloric acid produces chlorides and sulfuric acid produces sulfates. Students should be able to complete word equations for these reactions, given the names of the reactants.	5.4.2.1	4.7.3.1	Chemistry 2 1: Acids and alkalis	Happy Holi!
		The test for hydrogen uses a burning splint held at the open end of a test tube of the gas. Hydrogen burns rapidly with a pop sound.	5.8.2.1	4.7.3.1		
		Acids are neutralised by alkalis (e.g. sodium hydroxide) and bases (e.g. insoluble metal oxides) to produce salts and water.	5.4.2.2	4.7.3.2		
		Acids are neutralised by carbonates to produce salts, water and carbon dioxide. Students should be able to complete word equations for these reactions, given the names of the reactants. Carbon dioxide turns limewater milky.	5.4.2.2	4.7.3.1		
		Salt solutions can be crystallised to produce solid salts.	5.4.2.3	4.7.3.2		

3.4.2	Energy and rate of reaction	Some reactions transfer energy to the surroundings so the temperature increases. Such reactions include combustion, oxidation and neutralisation.	5.5.1.1	4.7.3.3	Chemistry 2 2: Chemistry and energy AND 3: Speeding up	Warming up! AND Cooling down AND Kill that acid!
		Other reactions take in energy from the surroundings, so the temperature decreases. These reactions include dissolving ammonium chloride in water and reacting citric acid with sodium hydrogencarbonate.	5.5.1.1	4.7.3.3		
		The rate of a chemical reaction may be increased by increasing the temperature, increasing the concentration of reactants, increasing the surface area of solid reactants or by adding a suitable catalyst.	5.6.1.2	4.7.4.6		
3.4.3	Earth's atmosphere	During the first billion years of the Earth's existence, there was intense volcanic activity that released gases that formed the early atmosphere and water vapour that condensed to form the oceans. The early atmosphere was mainly carbon dioxide with little or no oxygen.	5.9.1.2	4.4.1.1	Chemistry 2 5: Air and water	Fit to breathe?
		From about three billion years ago, algae and plants developed and produced the oxygen that is now in the atmosphere, by a process called photosynthesis. Photosynthesis can be represented by the word equation: carbon dioxide + water → glucose + oxygen	5.9.1.3	4.4.1.1		
		Carbon dioxide was removed from the early atmosphere by dissolving in the oceans and by photosynthesis. Most of the carbon from the carbon dioxide gradually became locked up in rocks as carbonates and fossil fuels.	5.9.1.4	4.4.1.1		
		The Earth's atmosphere is now about four fifths (80 %) nitrogen and about one-fifth (20 %) oxygen, with small amounts of other gases, including carbon dioxide, water vapour and argon, which is a noble gas.	5.9.1.1			

3.4.4	Fuels and human impacts on the atmosphere	Crude oil is a mixture of a very large number of compounds. Crude oil is found in deposits underground, e.g. the oil fields under the North Sea.	5.7.1.1	4.8.1.2	Chemistry 2 4: Fuel and fires	Big oil
		Crude oil may be separated into fractions by fractional distillation. This process, which takes place in a refinery, can be used to produce a range of useful fuels and oils, including petrol and diesel.	5.7.1.2	4.8.1.3		
		When fuels burn completely the gases released into the atmosphere include carbon dioxide, water (vapour), and oxides of nitrogen. Sulfur dioxide is also released if the fuel contains sulfur. When fuels burn in a limited supply of air carbon monoxide is produced. Solid particles (soot) may also be produced. Students may be required to describe the impact on the environment of burning fossil fuels. Oxides of nitrogen and sulfur dioxide cause acid rain and problems for human health. Carbon monoxide is a colourless, odourless, poisonous gas that can cause death. Solid particles can cause global dimming and problems for human health.	5.9.3.2	4.4.1.6	Chemistry 2 4: Fuel and fires AND Biology 1 3: Safe Earth	Up in flames AND Big stink AND Let's save the world!
		Some human activities increase the amounts of greenhouse gases in the atmosphere, such as carbon dioxide from burning fossil fuels and methane from landfill and cattle farming.	5.9.2.2	4.4.1.4	Chemistry 2 5: Air and water	The greenhouse effect
		Increased levels of greenhouse gases in the atmosphere cause the temperature to increase. Many scientists believe that this will result in global climate change.	5.9.2.3	4.4.1.4		
3.4.5	Water for drinking	Water that is safe to drink has small amounts of dissolved substances and low levels of microbes. Most drinking water is produced by choosing a suitable source of fresh water, filtering to remove solids and sterilising to kill microbes. If supplies of fresh water are limited, salty water can be distilled to produce fresh water. This requires a large energy input.	5.10.1.2	4.4.1.8		Fancy a drink?

3.5.1	Energy, energy transfers and energy resources	Describe, for common situations, the changes involved in the way energy is stored when a system changes. For example: • an object projected upwards • a moving object hitting an obstacle • a vehicle slowing down • bringing water to a boil in an electric kettle. Students may be required to describe the intended energy changes and the main energy wastages that occur in a range of devices.	6.1.1.1	4.1.1.4, 4.7.1.9 and 4.7.2.8	Physics 1 1: Full of energy	Saving Iron Man
		Energy can be transferred usefully, stored or dissipated, but cannot be created or destroyed.	6.1.2.1	4.8.2.6		
		The idea of efficiency. Whenever there are energy transfers in a system only part of the energy is usefully transferred. The rest of the energy is dissipated so that it is stored in less useful ways. This energy is often described as being 'wasted'. Unwanted energy transfers can be reduced in a number of ways, e.g. through lubrication and the use of thermal insulation.		4.8.2.6	Physics 1 1: Full of energy AND 2: Saving money	Saving Iron Man AND Efficiency AND Where does the energy go?
		How the rate of cooling of a building is affected by the thickness and thermal conductivity of its walls. The higher the thermal conductivity of a material the higher the rate of energy transfer by conduction across the material.	6.1.2.1	4.8.2.6	Physics 1 2: Saving money	Where does the energy go?
		Describe the main energy resources available for use on Earth. These include fossil fuels (coal, oil and gas), nuclear fuel, bio-fuel, wind, hydroelectricity, geothermal, the tides, the Sun, water waves. Distinguish between energy resources that are renewable and energy resources that are non-renewable.	6.1.3	4.8.2.4	Physics 1 3: Bright sparks!	Power stations

3.5.2	**Forces and work**	A force is a push or pull that acts on an object due to the interaction with another object. All forces between objects are either: • contact forces – the objects are physically touching • non-contact forces – the objects are physically separated.	6.5.1.2	4.6.1.1	Physics 1 4: The force	Pinball wizards
		When a force causes an object to move through a distance, work is done on the object.	6.5.2	4.6.1.3		
		Work done against the frictional forces acting on an object causes a rise in the temperature of the object.	6.5.2	4.7.1.10		
3.5.3	**Speed and stopping distances**	Speed is calculated from the distance travelled in a certain time. Units of speed include metres per second and kilometres per hour. Simple calculations of average speed using the equation: speed = distance/time will be required.	6.5.4.1.2	4.7.1.1	Physics 1 5: Road safety	Speed kills
		The stopping distance of a vehicle is the sum of the distance the vehicle travels during the driver's reaction time (thinking distance) and the distance it travels under the braking force (braking distance). For a given braking force the greater the speed of the vehicle, the greater the stopping distance.	6.5.4.3.1	4.7.1.10		
		Reaction times vary from person to person. Typical values range from 0.2 s to 0.9 s. Knowledge and understanding of methods used to measure human reaction times. Knowledge of how a driver's reaction time can be affected by tiredness, drugs and alcohol. Distractions may also affect a driver's ability to react.	6.5.4.3.2	4.2.1.6		
		The braking distance of a vehicle can be affected by adverse road and weather conditions, and poor condition of the vehicle. Students should be able to analyse a given situation to identify how braking could be affected.				

3.5.4	Atoms and nuclear radiation	Some atomic nuclei are unstable. The nucleus gives out ionising radiation. This is a random process called radioactive decay.	6.4.2.1	4.3.2.2	Physics 2 5: Death rays and diagnosis	Cook that pigeon!
		The nuclear radiation emitted may be: • alpha particles • beta particles • gamma rays. Properties of alpha particles, beta particles and gamma rays limited to their penetration through materials and their range in air. Students will be expected to know some of the uses and dangers of the three types of radiation.				
3.6.1	Electrical current	Electric current is a flow of electrical charge. The size of the electric current is the rate of flow of electrical charge.	6.2.1.2	4.7.2.1	Physics 2 1: Stage show	Wiring it up
		The current through a component depends on both the resistance of the component and the voltage across the component. The greater the resistance of the component the smaller the current for a given voltage across the component.	6.2.1.3	4.7.2.2		
		The resistance of a component is a measure of how difficult it is for an electric current to pass through it.	6.2.1.3	4.7.2.2		
		A complete circuit is necessary for a current to flow. Cells and batteries supply current that always passes in the same direction. This is called direct current (dc).	6.2.2	4.7.2.5		
		An alternating current (ac) is one that changes direction. Mains electricity is an ac supply. In the UK it has a frequency of 50 Hz and is 230 V.	6.2.3.1	4.7.2.5		
3.6.2	Domestic electricity	Most electrical appliances are connected to the mains using a three-core flex. The insulation covering each wire in the flex is colour-coded for easy identification: • live wire – brown • neutral wire – blue • earth wire – green and yellow stripes. The earth wire is a safety wire to stop the appliance becoming live and the fuse contains a thin piece of wire, which melts if the current becomes too large, thereby cutting off the supply. Students should be able to select the correct fuse from a list when given the current rating of an appliance. Some appliances do not have an earth wire because they are double insulated.	6.2.3.2	4.7.2.6	Physics 2 2: Power at home	The world's greatest plug

		Everyday electrical appliances are designed to bring about energy transfers. The amount of energy an appliance transfers depends on how long the appliance is switched on for and the power of the appliance.	6.2.4.2	4.7.2.8		
3.6.3	**Magnetism and electro-magnetism**	The poles of a magnet are the places where the magnetic forces are strongest. When two magnets are brought close together they exert a force on each other. Two like poles repel each other. Two unlike poles attract each other. Attraction and repulsion between two magnetic poles are examples of non-contact force. The patterns of magnetic fields between bar magnets will be required.	6.7.1.1	4.6.3.1	Physics 2 3: Magnets	Magnets in action
		When a current flows through a conducting wire a magnetic field is produced around the wire. The strength of the magnetic field depends on the current through the wire and the distance from the wire. Shaping a wire to form a solenoid increases the strength of the magnetic field created by a current through the wire. Adding an iron core increases the magnetic field strength of a solenoid. An electromagnet is a solenoid with an iron core. Students should be familiar with common uses of electromagnets, e.g. in scrapyard cranes and relays.	6.7.2.1	4.6.3.4		
3.6.4	**Different types of waves**	Waves may be either transverse or longitudinal. In a transverse wave the oscillations are perpendicular to the direction of energy transfer. The ripples on a water surface are an example of a transverse wave. In a longitudinal wave the oscillations are parallel to the direction of energy transfer. Longitudinal waves show areas of compression and rarefaction. Sound waves travelling through air are longitudinal.	6.6.1.1	4.1.4.1	Physics 2 4: Waves and wobbles	Sound it out

		Waves are described by their amplitude, wavelength and frequency.	6.6.1.2	4.1.4.2		
		The amplitude of a wave is the maximum displacement of a point on a wave away from its undisturbed position.				
		The wavelength of a wave is the distance from a point on one wave to the equivalent point on the adjacent wave.				
		Students may be required to use the equation:				
		wave speed (m/s) = frequency (Hz) x wavelength (m)				
		The frequency of a wave is the number of waves passing a point each second.				
3.6.5	Electromagnetic waves	Electromagnetic waves are transverse waves that transfer energy from the source of the waves to an absorber.	6.6.2.1	4.1.4.3	Physics 2 5: Death rays and diagnosis	Cook that pigeon
		Electromagnetic waves form a continuous spectrum and all types of electromagnetic wave travel at the same velocity through a vacuum (space) or air. The waves that form the electromagnetic spectrum are grouped in terms of their wavelength and their frequency. Going from long to short wavelength (or from low to high frequency) the groups are: • radio • microwave • infrared • visible light (red to violet) • ultraviolet • X-rays • gamma rays. Ultraviolet waves, X-rays and gamma rays can have hazardous effects on human body tissue. The effects depend on the type of radiation and the size of the dose.	6.6.2.1	4.1.4.3		

169

		Electromagnetic waves have many practical applications, e.g.:	6.6.2.4	4.1.4.3		
		• radio waves – television and radio (including Bluetooth) • microwaves – satellite communications, cooking food • infrared – electrical heaters, cooking food, infrared cameras • visible light – fibre optic communications • ultraviolet – energy efficient lamps, sun tanning • X-rays – medical imaging and treatments • gamma rays – for sterilising. Students should be able to give brief explanations of why each type of electromagnetic wave is suitable for the practical application.				

Glossary

Absorb Absorption is the taking in of one substance by another, e.g. a paper towel will absorb water.

Acid A chemical that turns blue litmus paper red – it can often dissolve things that water cannot.

Adaptations A feature of a living organism that helps it to survive, e.g. the zebra's stripes are an adaptation that makes it difficult to see in the wild and so protects it from lions.

Adrenal glands Glands that produce the hormone adrenaline in times of stress. They fit on top of the kidneys in humans.

Adrenalin A hormone produced by the adrenal gland which prepares the body for vigorous activity.

Algae A type of green plant that does not produce seeds or fruits. They are often found in water, seaweed is a good example of an alga. The plural for alga is algae.

Alkaline A substance which makes a solution that turns red litmus paper blue.

Amplitude The difference between the highest and lowest points on a wave, the larger the amplitude of a sound wave the louder the sound.

Antibiotics A substance produced by a microbe which kills other microbes. Some can be purified and used to treat infections.

Antibodies Chemicals produced by special cells called B lymphocytes which attack and disable invading micro-organisms.

Anus The ring of muscle at the end of the gut. It is normally squeezed shut to prevent waste leaving the gut.

Artificial selection Controlling the breeding of plants or animals by humans to emphasise useful features in their offspring.

Asexual reproduction Reproduction involving just one parent. The offspring are genetically identical to the parent.

Atmosphere The mixture of gases we call the air. The atmosphere is roughly 80 % nitrogen and almost 20 % oxygen, with other gases making up less than 1 %.

Atoms The smallest part of an element. Atoms consist of negatively-charged electrons flying around a positively-charged nucleus.

Attract Pulled together.

Bacteria These are microscopic single-celled living things that have no nucleus. Different bacteria can do everything from make us ill to make food taste better!

Base A chemical which produces an alkaline solution in water. Most metal oxides are bases.

Biodegradable Something is biodegradable if it can be broken down by living organisms in the environment. Paper is biodegradable because it rots but plastic is not because living organisms cannot break it down.

Blood donors People who give blood for others to use.

Boiling point The highest temperature a liquid can reach before it changes into a gas. The boiling point of water is 100°C.

Carbon A very important element. Carbon is present in all living things and forms a huge range of compounds with other elements. Coal and soot are almost pure carbon.

Carbon dioxide A gas containing only carbon and oxygen. Its chemical formula is CO_2. Carbon dioxide is produced by respiration and combustion.

Carbon monoxide A poisonous gas made when methane burns in a limited supply of oxygen. Its chemical formula is CO.

Carnivores An animal that hunts for and eats other animals.

Cells The smallest parts of a living thing. Cells from multicellular organisms have a nucleus and cytoplasm and a range of other parts. Cells of some types of microorganisms do not have a proper nucleus.

Change of state A change of state is a change from a gas to a liquid or back again or from a solid to a liquid or back.

Charge Electric charge can be positive or negative. The charge produces a range of effects including the flow of an electric current. An electric charge moving along a conductor creates a magnetic field.

Chloride Chemicals containing chlorine and one other element, e.g. sodium chloride.

Chromosomes Threadlike structures in cells which contain genetic material.

Circulatory system A system of tubes and a pump to move fluids around a body. The heart and blood vessels are the circulatory system in humans.

Climate The pattern of weather in an area.

Combustion The scientific name for burning. In combustion reactions fuels react with oxygen to form carbon dioxide and water, and give out heat and light.

Compete If two living things are trying to get the same thing and it is in short supply, they will compete with each other for the resource.

Composition The substances that something contains.

Compounds These are groups of atoms bound together, in fixed proportions, by chemical bonds. Compounds have different chemical and physical properties from the elements that form them.

Conductivity Movement of energy as heat (or electricity) through a substance without the substance moving.

Conductor A substance that will let heat or electricity pass through it. A bar of copper is a good example of a conductor.

Consumer An organism in an ecosystem that uses up organic matter produced by other organisms. All animals are consumers.

Contact force A force which needs to be in contact with the object it is acting on. Bicycle brakes use frictional force to slow down the moving wheel.

Contract To get shorter or smaller.

Corrodes A corrosive substance is able to eat away at something else.

Crest The top of a wave.

Decay When things break down over time.

Decompose The bodies of living things break down after they die to produce simpler chemicals which can be recycled. This is called decomposition.

Dioxide A compound containing two oxygen atoms in its chemical formula.

Dissipated To dissipate is to spread out, a smell will dissipate through fresh air or heat will dissipate into cooler areas.

DNA The molecule that carries the genetic information in animals and higher plants. DNA is short for Deoxyribose Nucleic Acid.

Donor A person who provides an organ for a transplant.

Double blind A drug test in which neither the person handing out the treatments nor the patients who are taking it know who is getting the test drug and who is getting the placebo.

Dual circulatory system A circulatory system where the blood passes through the heart twice in every complete circuit of the body. One pump drives it to the lungs and the other drives it to the body.

Efficient A measure of how much energy is transferred by a system in the intended way. The most efficient systems transfer most of the input energy into the chosen energy output. Inefficient systems waste a lot of the input energy and produce a wider range of output.

Eggs A special cell made by a female. When it joins with a sperm from a male it can grow into a new individual.

Electric current The flow of electrical charge through an electrical circuit.

Electromagnetic A coil of wire, often surrounding an iron bar, that produces a magnetic field when electricity flows though the wire.

Electromagnetic wave A transverse wave with electrical and magnetic properties which can pass through a vacuum.

Element A substance that cannot be split into anything simpler by chemical means. All the atoms of an element have the same atomic number although some may have different atomic masses.

Endocrine Endocrine glands secrete hormones directly into the blood. The hormones effect other body parts.

Energy Energy is the ability of a system to do something (work). We detect energy by the effect it has on the things round us, such as heating them up, moving them, etc.

Enzymes Chemicals found in living organisms that speed up the rate of a chemical reaction.

Equations A shorthand way to show a change. Equations may use words or chemical formulae.

Evolution The gradual change in living organisms over millions of years caused by random mutations and natural selection.

Food chain A sequence of organisms showing what an animal eats e.g. grass - rabbit - fox.

Food web All the food chains in a particular area shown on one diagram. It shows how many of them link together. Most organisms are part of more than one food chain.

Force A force is a push or pull which is able to change the velocity or shape of a body. Forces only exist between bodies. Every force that acts on a body causes an equal and opposite reaction from the body.

Fossil Preserved evidence of a dead animal or plant. Fossils can be body parts or evidence of activity like tracks, burrows, nests or teethmarks.

Fossil fuels A fuel like coal, oil and natural gas formed by the decay of dead living things over millions of years.

Frequency The number of vibrations per second. Frequency is measured in Hertz.

Fuel This is something that gives out energy, usually as light and heat, when it burns.

Fuses A special component in an electrical circuit. A fuse contains a thin wire which is designed to heat up and melt if too much current flows through it. This breaks the circuit and the electricity stops flowing, which stops the current from damaging other, more valuable parts of the device.

Galena Galena is an ore that contains lead.

Gametes Special cells that join to form a new individual during sexual reproduction.

Gases All substances are solids, liquids or gases. A gas has no fixed shape and will expand to fill all of the space available.

Gene A gene is a length of DNA that tells the growing cells how to make particular chemicals. Genes help to determine the eventual structure and behaviour of an organism.

Glands Parts of the body that produce a secretion which has an effect outside the gland.

Glossary

Global warming The gradual rise in average global temperature over the last century or so. It is almost certainly caused by human activity - mainly burning of fossil fuels like coal, gas and oil.

Glucose A type of sugar. Glucose is sometimes called dextrose.

Gonads The sex organs – in male humans these are the testicles and in female humans they are the ovaries.

Greenhouse gases Gases like carbon dioxide, methane and water vapour that increase the greenhouse effect.

Groups Scientists say that all elements in the same column in the periodic table are in the same group. They have similar chemical and physical properties.

Gut The long tube that starts in your mouth and goes down to your bottom. It is sometimes called the digestive tract.

Halogens Elements from Group VII in the periodic table. Chlorine and bromine are both halogens.

Herbivores Animals that eat plants.

Hydrocarbons Hydrocarbon molecules are molecules that contain only carbon and hydrogen atoms. Many fuels are hydrocarbons, e.g. natural gas (methane) and petrol (a complex mixture).

Immune If you are immune to an illness you do not suffer from it.

Immune system The parts of the body that protect against illnesses. The lymph glands are particularly important in the immune system.

Indicators A chemical that changes colour in acid and alkaline solutions. Indicators are used to tell us the pH of a solution.

Inefficient In an inefficient energy transfer most of the energy is not transferred usefully - most is water in some other form.

Insulator A substance that will not let heat or electricity pass.

Longitudinal In longitudinal waves, the vibration is along the direction in which the wave travels.

Lubricant Oils or powders that help surfaces to slide over each other without friction.

Magnetic field An area where a magnetic force can be felt.

Magnetic field lines The areas in a magnetic field where the field strength is the same.

Magnet A magnet will attract objects made of steel and some other metals. Magnets have two poles: North and South.

Malignant A malignant tumour is one whose cells can spread into other tissues and cause more tumours.

Mate A mate is a sexual partner. Animals often compete for mates.

Medium The name given to the material that electromagnetic radiation is passing through.

Melting point The temperature at which a solid changes into a liquid or a liquid changes into a solid.

Menstruation The release of blood and cells from the body of a female when the wall of the uterus breaks down. Menstruation occurs only if the female is not pregnant.

Metals Metals are elements that are shiny, strong and conduct heat and electricity well. Most elements are metals and are found on the left hand side of the periodic table.

Microscopes A device for looking at very small objects.

Mineral Natural solid materials with a fixed chemical composition and structure. Rocks are made of collections of minerals. Mineral nutrients in our diet are things like calcium and iron. They are simple chemicals needed for health.

Mixture A mixture contains at least two different substances. Some mixtures are easy to separate but others are very difficult to split up into their different components. Most things in the world are mixtures.

Monomers The small molecules which join together to form a polymer molecule.

Mutation Variations caused by a change to an organism's genetic material.

Natural selection Factors in the environment affect animals and plants so that some survive to reproduce successfully and pass on their good combinations of genes. Others survive less well and do not pass on their poor combinations of genes as often. This is called natural selection.

Neutralisation A reaction between an acid and an alkali to produce a neutral solution.

Non-contact force A force which can act without touching the object it is acting on. Magnetism is an example of a non-contact force because the magnets do not have to touch to effect each other.

Nonmetals Nonmetals are elements that are not metals. They are more varied than metals and are not shiny, nor do they conduct electricity or heat well. They are found on the right hand side of the periodic table.

Non renewable energy sources Energy sources that will run out because they are not being made as quickly as humans are using them up. Oil, coal and natural gas are all non-renewable energy sources.

Obese An obese person is very fat. Their extra weight puts a strain on the heart and other body parts.

Offspring The organisms produced by reproduction.

Oil refineries A place where crude oil is separated into simpler chemicals and these are made into useful products like petrol or jet fuel.

Ore A mineral containing a useful level of a metal-containing compound or compounds.

Organ Organs are parts of the body that are self-contained and have a particular job to do.

Ovaries The part of the female body that produces the egg.

Oxides A compound containing oxygen and one other element, for example carbon dioxide contains only carbon and oxygen.

Oxidised A substance has been oxidised when oxygen has reacted with it.

Oxygen A colourless gas with no smell that makes up about 20 % of the air. All animals need a constant supply of oxygen to stay alive.

Pathogens Micro-organisms that cause diseases.

Penicillin A medicine that kills some dangerous micro-organisms. It is made from a type of fungus.

Period A vertical row in the periodic table.

Periodic table A chart showing the relationships between the elements based on their atomic number.

pH scale The range of levels of acidity or alkalinity. A pH of 7 is neutral. A pH below 7 is acid and the lower it goes the more acid it becomes. A pH above 7 is alkaline.

Photosynthesis The production in green plants of sugar and oxygen from carbon dioxide and water using light as an external energy source.

Placebo A treatment with no active ingredient used in drug trials.

Poles The parts of a magnet where the magnetic field is strongest. Poles can be either North or South.

Pollutant A chemical made by human activity that is damaging to the environment.

Polymer Single large molecules formed by joining together monomers.

Prism A transparent glass or plastic block with straight sides.

Producer Living organisms that use sunlight energy to build up organic matter. All primary producers are plants.

Property The things that are true about a substance. Important properties include the melting and boiling points, density and strength.

Puberty A time when hormones produce rapid growth of the body and the development of secondary sexual characteristics.

Pure Something that contains only one substance. So gold is pure gold and deionised water is pure water.

React A change in a group of chemicals where bonds between atoms are broken and remade to form new chemicals.

Reflex An inborn stimulus-response pair that is usually to do with protecting the body from harm. Reflexes are not usually under the control of the conscious brain.

Renewable A renewable resource is replaced as quickly as it is used. Wind and solar power both depend on renewable energy sources.

Repel To push apart, usually used to talk about magnetic repulsion which pushes like magnetic poles apart.

Reproduction Producing offspring. All living things must reproduce.

Resistance A force which tends to oppose movement through a medium, e.g. parachutists depend on the drag produced by their parachute to slow them down before they hit the ground is an example of air resistance. Wires carrying electric current also have resistance - it reduces the flow of electric current.

Respiration The chemical process that makes energy from food in the cells in your body. All living things must respire.

Rust A complex mixture of chemicals made when iron reacts with oxygen and moisture. It is often brown in colour.

Salt A compound made when an acid reacts with an alkali.

Scrotum The bag of skin that hangs outside the body cavity and contains the testicles in males.

Sexual reproduction Reproduction that involves two sexes producing specialised cells, the gametes, which join together to produce the next generation. Sexual reproduction produces offspring that are different from either of the parents.

Side effects An effect that happens but is not the main effect of the treatment. Drugs can sometimes have side effects which do not always help the body even though the main effect is to cure the illness.

Sodium A soft, reactive Group 1 metal.

Sodium chloride The scientific name for common table salt.

Sperm Special cells produced by the male. A sperm from a male joins with an egg from a female to produce a baby.

State The state of a substance is whether it is a solid, liquid or gas.

Substance Something which has mass and a fixed composition. Water is a substance. Gold is a substance.

Survival of the fittest Darwin's idea that the offspring of those animals and plants which are most suited (fittest) to their environment are most likely to survive into the next generation.

System Organs are gathered together into systems to carry out a major function in a body. So, the circulatory system includes The heart and blood vessels that work together to move substances around the body.

Target organ The organs which are affected by a particular hormone, e.g. the ovaries are the target organs for the hormone FSH which is secreted by the pituitary gland.

Testicles The male sex organ which produces the male gametes (sperm).

Thermal conductivity The speed at which heat can move through a material. Metals have high thermal conductivity because they allow heat to move through easily.

Tissue A group of cells of the same type, so nervous tissue contains only nerve cells.

Top carnivore The animal that eats other animals but is not eaten by anything else.

Transplant A transplant is an operation that moves an organ from one person (the donor) to another.

Transverse In transverse waves, the vibration is at right angles to the direction in which the wave travels.

Trough The lowest point in a waveform.

Tumour A mass of tissue formed by the uncontrolled division of cells. These cells cannot carry out normal cell functions and may damage healthy cells by pushing against them.

Uterus The organ in the female where the baby grows during pregnancy. Also known as the womb.

Vaccination A solution of weak or dead microbial components that is injected into a person. It stimulates the immune system but does not cause the disease.

Variation The existence of a range of individuals of the same group with different characteristics.

Viruses A tiny micro-organism that lives inside other cells. Some viruses cause very serious diseases like smallpox.

Volt The international unit of electrical potential. The mains voltage in the UK is 240 Volts.

Wave Waves transfer energy across large distances by making things move very slightly back and forth. A small movement in one place will start a movement in the place next to it, and so on.

Wavelength The distance from one place on a waveform to the same place on the next waveform.

White blood cells A group of blood cells produced by the bone marrow and lymph glands which are involved in the immune response.

Word equations Word equations summarise a chemical reaction in words.

Work Work is done when a force makes something move.

Index

Index

Index